Plastics Gearing

MECHANICAL ENGINEERING

A Series of Textbooks and Reference Books

EDITORS

L. L. FAULKNER

Department of Mechanical Engineering
The Ohio State University
Columbus, Ohio

S. B. MENKES

Department of Mechanical Engineering
The City College of the
City University of New York
New York, New York

ADDITIONAL VOLUMES IN PREPARATION

Mechanical Engineering Software

Plastics Gearing
SELECTION AND APPLICATION

CLIFFORD E. ADAMS

IBM Corporation
Charlotte, North Carolina

MARCEL DEKKER, INC.   New York and Basel

To my wife of 35 years, Gladdy, a woman of cheery disposition, unending encouragement, and unusual patience

Library of Congress Cataloging-in-Publication Data

Adams, Clifford E., [date]
    Plastics gearing.

    (Mechanical engineering ; 49)
    Includes index.
    1. Plastic gearing. I. Title. II. Series.
TJ184.A33   1986   621.8'33   85-31146
ISBN 0-8247-7498-1

Marcel Dekker, Inc.
270 Madison Avenue, New York, New York 10016

Current printing (last digit):
10 9 8 7 6 5 4 3 2 1

Printed in the United States of America

Preface

This book has been written primarily for the design engineer and technologist. The user whose daily effort is not strictly gearing- or plastics-oriented should find here the help needed to accomplish a reliable plastics gearing system design. The book may also be useful in preparing technical students in mechanical design, or as an initiation into plastics applications of a specialized nature. The experienced gear engineer may also find the plastics information valuable, depending on the depth of his daily involvement with plastics gears.

It will be helpful if the reader has some knowledge of gearing and of plastics. The material is presented in concise terms and form, so the most essential formulas, data, and information will unite the gearing and plastics disciplines in a readily understood and useful manner. Many texts, handbooks, standards, research papers, design catalogs, data sheets, technical magazine articles, and mathematical tables are available on both gearing and plastics. It is usually necessary first to conduct an exhaustive literature search and gather all the readily available information, then to extract applicable design, engineering, and application data. Often some important piece of information or data is overlooked or its significance misunderstood. This book eliminates the need for such time-consuming and questionable procedures. The progression through various steps in the text shows how formulas, illustrations, and tables are to be used in actual design situations. The uninitiated or occasional gearing designer should find this method helpful in preventing design pitfalls.

There are basically two methods of proceeding with plastics gearing selection and application. The first is to purchase plastics gears from standard stock gear suppliers. A number of companies provide catalogs to designers listing readily available parts. Design time can be reduced using this useful approach. However, there are some key factors to consider. The technician or designer must be thoroughly aware of all his application demands. He must also study all the engineering data provided to be sure that the correct parts are ordered.

The second method is to design the gearing completely and provide all the specifications for manufacturing and assembly. Phases of design that must be adjusted from metallic to plastics principles are discussed in this volume. Materials and material selection and their relationship to the ever-important gear rating are also presented. Included are such vital topics as tooth variations, tolerances, formats, and inspection of plastics gears. In the application of plastics gearing, the focus is on the output accuracies, mountings, lubrication, testing, gear fastening, and other aspects that contribute to satisfactory product life. Where tables will assist the designer, they are included for efficient and convenient use.

In such an undertaking as the writing of this book, it is impossible to include the entire repertoire of the plastics gearing engineer. It is also impossible to avoid differences of opinion in areas where general consensus does not now exist. Where there are these differences of opinion, data is presented that has proven successful in the author's plastics gearing applications. A further benefit of this book is that the stimulation of ideas that will take place within plastics, gearing, industrial, academic, or standards organizations will perhaps inspire efforts to resolve problems in areas where more research is needed.

Acknowledgments

No small amount of confusion and frustration has led to the gathering of published data on gearing and plastics over the past 23 years. The enormous amount of data available only added to the problem of providing IBM design engineers with a unified design system. Obviously, a book of this nature would not be possible without the direct and indirect support and cooperation of a large number of individuals and organizations. My work with IBM designing, analyzing, and providing engineering design approval on mechanical drive systems provided the opportunity for significant contacts, discussions, joint activities, and resolution of technical problems.

I owe gratitude, first of all, to IBM for its support and encouragement without which this task would not have been undertaken. Also, the American Gear Manufacturers Association (AGMA) has been particularly helpful in providing information, contacts, and a suitable forum for the expression of gearing concerns in all phases of the discipline. This association has permitted a generous use of AGMA standards, information sheets, and publications. Plastics information has been liberally provided by the members of the Society of Plastics Engineers, and illustrations by the American Society for Metals. The willing assistance and contributions of the many authors and publishing companies, too numerous to mention here, can be discerned throughout the text. Also gratefully acknowledged are plastics and gear producers, material suppliers, and stock gear producers who have been especially helpful.

I also wish to express appreciation to my IBM management and colleagues, Richard F. Smith, David G. Sedor, Richard F. Tynan, David A. Opp, James P. Aiken, and Robert P. Leo. I must thank many of my peers associated with AGMA, Donald L. Borden, Paul M. Dean, Jr., Darle W. Dudley, F. Reed Estabrook, Jr., William W. Ingraham, Edward J. Wellauer, and many others, not only for assistance with the book, but for their profound influence in the gearing and plastics fields as well as in my professional life. I especially appreciate the interest, helpful attitude, and thoughtful comments of the manuscript reviewers.

A note of gratitude is also expressed to the American Red Cross, the North Carolina National Guard, and the fine folks that manned the Brunswick County Evacuation Center at the Shallotte, North Carolina, high school. Shelter, food, tables to write on, and chairs to sleep on were provided during the passing of hurricane Diana in September 1984. The rough drafts of Chapters 7 through 12 were written during this 36-hour period.

A special note of thanks is due to John P. Dombrowski who became my mentor at IBM when I began my career in the gearing field in 1962. His personal instruction and direction were characterized by a remarkable patience and resilience.

Clifford E. Adams

Contents

Conversion Factors

To convert from	To	Multiply by
Foot/sec^2	Meter/sec^2 (m/s^2)	3.048×10^{-1}
Inch/sec^2	Meter/sec^2 (m/s^2)	2.54×10^{-2}
Kg-force-meter	Newton meter (N·m)	9.806650
Ounce-force-inch	Newton meter (N·m)	7.061552×10^{-3}
Pound-force-inch	Newton meter (N·m)	1.129848×10^{-1}
Pound-force-foot	Newton meter (N·m)	1.355818
Pound-force-foot/inch	Newton meter/meter (N·m/m)	5.337866×10^{1}
Pound-force-inch/inch	Newton meter/meter (N·m/m)	4.448222
Kilogram-force	Newton (N)	9.806650
Ounce-force (Avoirdupois)	Newton (N)	2.780139×10^{-1}
Pound-force (Avoirdupois)	Newton (N)	4.448222
Pound-force/inch	Newton/meter (N/m)	1.751268×10^{2}
Pound-force/foot	Newton/meter (N/m)	1.459390×10^{1}
Foot	Meter (m)	3.048×10^{-1}

(continued)

To convert from	To	Multiply by
Inch	Meter (m)	2.54×10^{-2}
Microinch	Meter (m)	2.54×10^{-8}
Micron (micrometer)	Meter (m)	1.0×10^{-6}
Mil	Meter (m)	2.54×10^{-5}
Newton/square meter	Pascal (Pa)	1.0
Pound-force/square foot	Pascal (Pa)	4.788026×10^{1}
Pound-force/square inch	Pascal (Pa)	6.894757×10^{3}
Degree Fahrenheit	Degree Celsius (°C)	$t_C = (t_F - 32)/1.8$
Foot/minute	Meter/second (m/s)	5.08×10^{-3}
Psi	Pascal (Pa)	6.894757×10^{3}
Module	Diametral pitch (P, Pd)	25.4

Introduction

Several plastics have been used successfully in gearing applications. The use of plastics gears has prompted many studies in both the plastics and gear industries. This book will discuss some of the more important aspects of plastic gears as examined by both disciplines. The discussion will include selection of gears that are purchased from companies manufacturing to recognized standard dimensions and that may be ordered from the manufacturer's stock. Without adequate knowledge on the user's part, the gears can be misapplied. The designer may choose to design his or her own gearing. Whichever approach is taken will determine key steps to be followed by the gear engineer or designer in the design process.

The first consideration will be the selection process of plastic gearing, including the basic design criteria. Enough material is provided so that the most common gears can be selected without reference to other sources.

A note of caution is necessary in regard to the use of the American Gear Manufacturers Association (AGMA) publications. With the exception of AGMA 141.01, they are developed for metal gears. The basic concepts of gearing do not differ except in the material property differences and other areas noted in this book. A reference listing and other notations are given to AGMA standards, design handbooks, and data sheets. A check with AGMA should be made to insure that the information used is the most current data available. Generally, the tables and formulae are at this time well established, and changes are relatively infrequent. In reference to plastics, on-going investigations and development of new materials, material processing, tool and mold design, and application studies are actively underway. The material covers what is known today and that which has been applied

successfully. The Society of Plastics Engineers (SPE) has sponsored
some recent publications that are also used for reference.

No attempt is made to include all of the vast information accumulated
over the years in gearing or in plastics. Fairly comprehensive references are provided for guidance in in-depth gearing studies. Also,
additional information is available from the manufacturers and proces-
sors of both plastics and gearing.

Gears are mechanical elements that transmit power or angular motion by successively intermeshing teeth. The spur, the most common
type, operates between two parallel shafts. Most other types can be
considered adaptations of the spur gear, and special calculations are
usually required for each. Although our discussion of plastic gearing
will be centered on spur gears, the application to other types is similar. The gear that is selected is usually determined by the position
of the gear axes or shaft arrangement — parallel, intersecting, or
nonparallel—nonintersecting. Special treatment must also usually be
given certain types of gears.

The following is a list of advantages and disadvantages of plastics
gearing. Of course, not all plastics can provide all the advantages,
nor are they handicapped by their disadvantages. Just as in metal
gearing, both the ultimate success and total life of the gear set are
influenced by the completeness of the designer's knowledge, the mag-
nitudes of all the variables, and the allowances provided for each vari-
able in the final design. For example, many of the advantages relate
to injection-molded gears rather than to cut gears. A realistic balance
between the advantages and disadvantages will provide the user with a
satisfactory gear design and final product.

Advantages:

    Relative low cost
    Ability to dampen moderate shock and impact
    Ease and speed of manufacture
    Elimination of finishing, plating, and machining steps
    Inherent lubricity; low coefficient of friction
    Less critical tolerances
    Lightweight; low inertia
    Little or no lubrication necessary
    Noise reduction
    Resistance to abrasion, chemicals, corrosion, creep, and vibration
    Smooth, quiet performance
    Wide range of configurations and complex shapes possible

Disadvantages:

    Ambient temperature and temperature at tooth contact surface must
       be limited

Machining considerations; cost for molds to produce correct tooth
  forms
Maximum load-carrying capacity decreases as temperature increases
Maximum load-carrying capacity lower than metal gears
May be affected by lubricants, solvents, certain chemicals, and
  humidity
Sections may provide stress raisers
Susceptible to moisture absorption
High thermal coefficient of expansion

Well before the start of design, a thorough study must be made to
determine the conditions under which the gears must operate.   This
study must consider the gears as part of the complete system. Shafts,
bearings, mountings, lubrication, couplings, enclosures, seals, and
the prime mover are all part of the system and must be included in
the study.   The objective is to define the total unit in its operating
mode and environment for its anticipated life.   Failure of one part of
the system can lead to catastrophic system failure.   Characteristics
of the prime mover and the required accuracy, efficiency, and life of
each component must be compatible.   Therefore, the following items
should be known or calculated.

1.   Prime mover torque or horsepower and its characteristics, such
     as overload potential, peak loads, and their duration, accelera-
     tion or deceleration inertia loads, and uniformity of the output
     torque
2.   Angular accuracy of the output where the accuracy of angular
     motion is important and loads are light
3.   Center distance and permissible center distance tolerances
4.   Gear ratio based on input/output speed
5.   Speed, as a heat-generating source, and ambient temperature
6.   Space limitation
7.   Environment and contamination possibilities
8.   Type and method of lubrication, if lubrication is used
9.   Protection of manufacturers, assemblers, and operators
10. Expected life of the unit or machine

Basic equations for plastic gears are the same as those used for
metal gears, although modifications to improve the operating characteristics or allow certain manufacturing processes are often recommended.   The approach chosen here, discussion of the selection of
plastic gears and then their application, will necessitate reference
among sections.   All sections will be tied together to illustrate use of
the design methods recommended.

1

Gear Tooth Action

Gears are known as positive motion elements because they transmit power or angular motion by successively intermeshing teeth that prevent skipping of teeth or slippage of the power transmission or motion. In the past, plastics gears have been thought of as being unable to transmit power, limited as to top operating speeds, and unable to transmit motion to a high degree accurately. Such blanket indictments are not valid in today's plastics gearing. Developments in the plastics industry have resulted in the availability of materials that can carry higher loads than were previously thought to be possible. The use of plastics gears in some power transmission applications has resulted in improved design of related equipment and of the overall mechanical system efficiency.

A mere increase in the factor of safety will not always guarantee a satisfactory design. In plastics gearing, designs used for power transmission, transient loads, ambient and surface temperatures, and tooth action should be known and allowed for in regard to material selection. More teeth in mesh, a pitch change, or wider tooth face may or may not be the correct approach. Tooth root and tip modifications, selection of a different material or mating materials, selection of a different motor, or bearing change may provide an adequate power transmission system using plastics gears. In this regard, a careful consideration of the gear lubrication system may indicate that available effort should provide cooling of the gear tooth surfaces as a viable design improvement.

Power Transmission

The important consideration in plastics power gearing is to have a
thorough knowledge of the whole range of possible power characteristics and design for every eventuality.

Operating speed of plastic gear sets is directly related to heat generation in many cases that lowers the plastics material properties and
results in gear failure. When running plastics gears at high speeds,
some design approaches have been to

1. Mate a plastic gear with a metal pinion.
2. Design the plastic gear with a metal insert.
3. Design the gear so plastic teeth are molded on a metal disk.
4. Provide air flow directly onto the gear mesh.
5. Provide lubricant as a coolant.
6. Select material or material combinations that are less susceptible
    to heating problems.
7. Limit the maximum speed or time of rotation for any one duty cycle.
8. Design for use of as fine a gear pitch as possible since sliding ve-
    locities are lower for smaller teeth.
9. Use one or more of the above in combination.

Transmission of Motion

Accurate transmission of motion is thought to be associated mostly with
fine-pitch gears and small gears that are usually ideal for production
by molding. The American Gear Manufacturers Association (AGMA)
quality classification system provides a well-defined means of specifying allowed tolerance values (1). Therefore, if a gear is held to a
quality class of Q10, tolerance values for the AGMA quality class Q10
apply regardless of the gear material or the production method. It
then follows that if gears to an AGMA quality class Q10 are used in a
system with other components to a like degree of accuracy, the transmission of motion will be the same for plastics gears as for metallic
gears. The limiting factor seems to be determined by the ability to
produce gears to the desired quality class and the stability of the material after shrinkage and in the operating environment. A proper
mating of gear materials and quality class, component class, physical configuration, and tolerances can produce a mechanical system
capable of a high degree of reliable transfer of motion.

Power and motion transmission can only be determined in the final
product, but attention to design variables can provide a reasonably
accurate prediction, assuming that design, manufacturing, and assembly specifications are followed. A comprehensive treatise on power
transmission losses describes in detail procedures for dealing with the

TABLE 1.1 Gear Dimensions and Control Criteria

1. Functional Gear Element Dimensions and Criteria (have a direct and sensitive effect on gear performance).

   Tooth form dimensions: addendum, dedendum, clearance, pressure angle
   Pitch: tooth-to-tooth composite error
   Tooth thickness: position error
   Tooth profile form: outside diameter and runout
   Lead angle: surface roughness of active profile
   Total composite error: material, hardness, and finish coating

2. Functional Nongear Element Dimensions and Criteria (not associated with gear teeth but affect performance).

   Mounting bore or journal diameter
   Runout between tooth elements and mounting diameter
   Lateral runout of mounting or registering surface
   Convexity, concavity, and parallelism of mounting surfaces

3. Nonfunctional Dimensions and Criteria (relate to the gear body but do not affect function).

   Hub design dimensions
   Web design dimensions
   Fastening design dimensions: keyway, pin hole, set screw, split hub, clamp, couplings

subject (2). Plastics gearing design will usually be concerned with gear mesh friction, bearings both plastic journal and ball, and dynamic loads. Since power transmission using plastics gears has been fairly uncommon, dynamic effects are often overlooked in the design stage. With the emergence of improved load-carrying-capacity materials the following should be considered in power gearing.

   Tooth loading and deflection
   Tooth spacing and composite errors
   Pitch line runout
   Elasticity of gear materials and mountings
   Torsional deflection and beam bending of the shafts
   Mountings and alignments
   Temperature differentials

Plastics gears are found today in both commercial and precision type applications for transfer of motion. Gear dimensions and control criteria have been presented by Michalec in three descriptive categories and are presented in Table 1.1. Fastening design dimensions are listed with items that do not affect function. This assumes that clearances and dimensions are such that fasteners do not cause the gearing to be mounted in an eccentric condition.

References

1. *AGMA Gear Handbook, Volume 1: Gear Classification, Materials, and Measuring Methods of Unassembled Gears* (AGMA 390.03), American Gear Manufacturers Association, Alexandria, VA (1973).

2. G. W. Michalec, *Precision Gearing Theory and Practice*, John Wiley and Sons, New York (1966).

2

Gear Types and Arrangements

Of the many types of gears available, they are grouped here according to the mounting shaft arrangement in Table 2.1. Straight-sided
and helical splines and timing belt drive pulleys could also be included with parallel shaft gears, since they are basically straight-
toothed elements with similar design principles. Nearly all of the
listed gear types can be molded and all can be cut from plastics materials. Because molding is able to produce many complex shapes,
common practice is to combine gears with other components in the
same mold cavity. If a mold line is not detrimental to the operation
of a part, split molds can be used providing economies warrant the
manufacturing savings.

Table 2.1 is not all-inclusive of the gear types available nor does
it include all of the types that can be molded. These are the most
common type gears and whenever a new type is being considered,
both gear and mold suppliers should be consulted for assistance.
As a guide, the nominal range of reduction ratios are listed. Ratios
of 300:1 or higher are possible where the upper ratio of 100:1 is
shown (1). The ratio of 100:1 is shown as a normal maximum limit.
(See also Figures 2.1 through 2.21.)

Spur Gear Recommendation

Gear arrangements for plastics gears follow the same basic guidelines
as for metallic gears except for certain specific areas that will be discussed. The most common gear arrangements are the parallel axis
external spur and then the helical gear. Metallic spur gears are
usually recommended for transmission of angular motion to parallel

TABLE 2.1 Mounting Shaft and Gear Type Relationship and Nominal Range of Reduction Ratio

---

Parallel Shafts

    Spur: external and internal (1:1-5:1) and (1:5-7:1)
    Helical: external and internal (1:1-10:1) and (1.5:1-10:1)
    Internal herringbone or double helical (2:1-20:1)
    External herringbone or double helical gear (1:1-20:1)

Intersecting Shafts

    Bevel: straight, spiral, Zerol®, Beveloid® (1:1-8:1)
    Face gears (3:1-8:1)

Nonparallel—Nonintersecting Shafts

    Crossed helical (1:1-100:1)
    Worms: single and double-enveloping (3:1-100:1)
    Face gears (3:1-8:1)
    Beveloid® (1:1-100:1)
    Helicon® (3:1-100:1)
    Hypoid (1:1-10:1)
    Spiroid® (9:1-100:1)

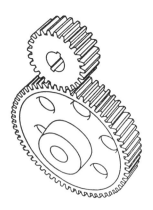

FIGURE 2.1 Spur gears. (From Ref. 5.)

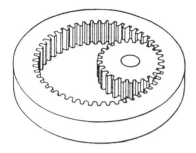

FIGURE 2.2 Internal spur gear set. (From Ref. 5.)

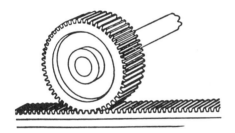

FIGURE 2.3 Spur rack and pinion. (From Ref. 5.)

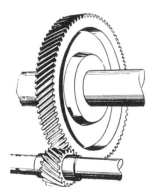

FIGURE 2.4 Helical gears. (From Ref. 5.)

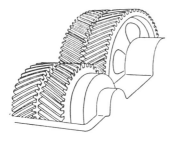

FIGURE 2.5 Double helical gears. (From Ref. 5.)

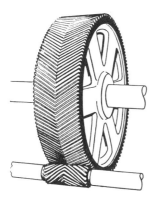

FIGURE 2.6 Herringbone gears. (From Ref. 5.)

FIGURE 2.7 Straight bevel gears. (From Ref. 5.)

FIGURE 2.8 Spiral bevel gears. (From Ref. 5.)

FIGURE 2.9 Zerol bevel gears. (From Ref. 5.)

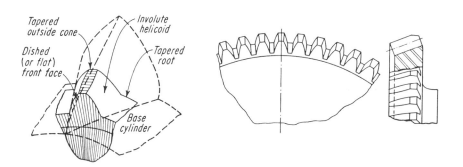

FIGURE 2.10 Beveloid tooth form. (From Ref. 1.)

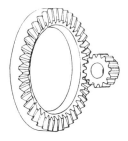

FIGURE 2.11 Face gear set. (From Ref. 5.)

FIGURE 2.12 Crossed-axis helical gears. (From Ref. 5.)

FIGURE 2.13 Cylindrical worm. (From Ref. 5.)

FIGURE 2.14 Single-enveloping worm gear set. (From Ref. 5.)

shafts under all load conditions at speeds up to 1000 RPM or 1000 feet per minute pitch line velocity, where maximum smoothness and quietness are not of prime importance. For plastics spur gears the maximum allowable speed or pitch line velocity must be adjusted according to the material used and method of design that will dissipate the tooth heat generated. Noise generation of many plastics materials is usually less than that generated by an equivalent set of metallic spur gears. It must be emphasized that specific velocity values are only recommendations, since material type and ability to control heat build-up play a large part in limiting or permitting increased velocity.

The advantages of spur gear sets over helical gears are their adaptability to timing applications involving reversal of motion or over-running loads. Since no axial thrust is produced, simplified shaft bearing design is a positive factor in holding backlash to a minimum. Also, spur gear teeth are less expensive to cut when producing electrodes for mold production, and inspection master gears are standard for all involute gears of the same pitch.

FIGURE 2.15 Hour-glass worm. (From Ref. 5.)

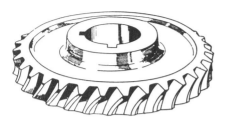

FIGURE 2.16 Double-enveloping worm gear. (From Ref. 5.)

    The only basic disadvantage in the use of spur gears is the lim-
ited tooth-to-tooth overlap that results in some vibration and noise
even at relatively low speeds. Once again, these disadvantages are
most pronounced in metallic gears. Spur gear noise of metallic sets
may be reduced by mating a plastic gear with a metal gear or using
a plastic gear set. A reduction in allowable power transmission may
be involved, so the design must take into consideration this aspect.
Just as metallic spur gears have advantages, some of the same benefits and others are possible with plastics spur gears.

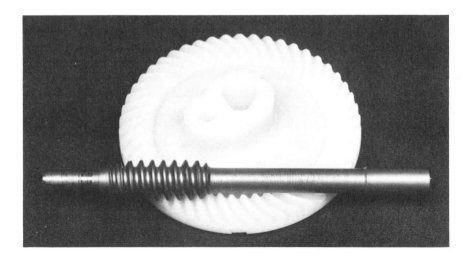

FIGURE 2.17 Helicon gear set. (Courtesy of ITW Spiroid, Chicago,
IL.)

FIGURE 2.18 Hypoid gear set. (From Ref. 5.)

Helical Gear Recommendation

Helical gears should be used for transmission of angular motion to
parallel shafts under all load conditions at speeds exceeding those
acceptable for spur gears for the particular material used. They
are also recommended for low speed applications where smooth and
quiet operation are important. Helical gears can also be used to
transmit angular motion to nonintersecting angular shafts, but their
load-carrying capacity is extremely limited in this type of application. Helical gears in this arrangement are usually called crossed-
helical gears, and the load limitation is due to tooth point contact
as opposed to spur and helical tooth line contact. Worms can also
be mated with helical gears instead of throated worm gears to obtain
high ratios under light load conditions, where backlash due to wear
is not functionally important.

The principal advantage gained in the use of helical gears over
spur gears is the increased overlap in tooth action by which the load
is transferred gradually and uniformly through a combined sliding
and rolling contact as successive teeth come into engagement. This
tooth action results in smooth, quiet operation even at high speeds.
Also the dampening characteristic of the plastics material itself aids
in reducing operational sound levels. An interesting set of recommendations (2) is summarized as follows.

Sound Level of Gearing

To make gears as quiet as possible

1. Specify the finest pitch allowable for load conditions.
2. Use a coarser pitch to produce a lower pitch sound.

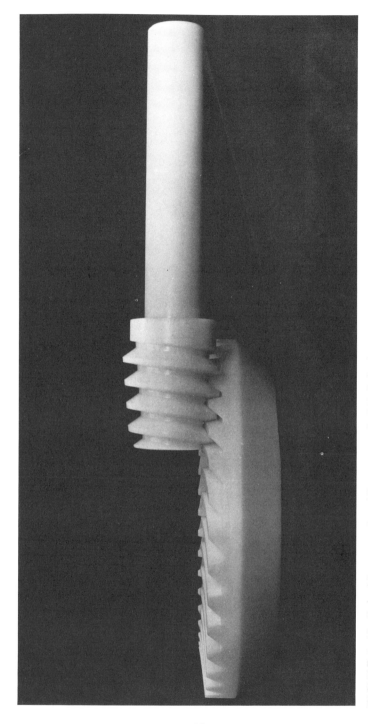

FIGURE 2.19 Spiroid gear set. (Courtesy of ITW Spiroid, Chicago, IL.)

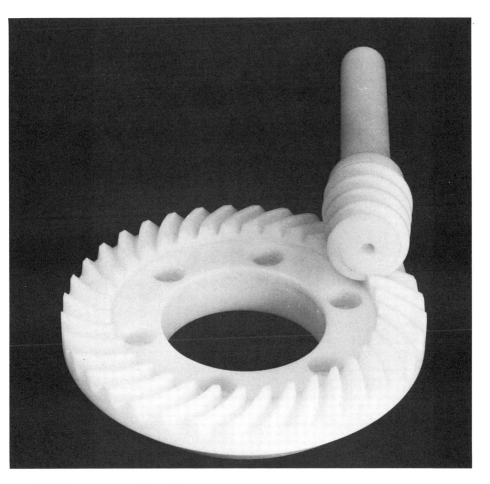

FIGURE 2.19 (continued)

3. Use a low-pressure angle.
4. Use a modified profile with root and tip relief.
5. Provide increased backlash.
6. Use high quality gears with a surface finish of 20 μin. or better.
7. Balance the gear set.
8. Use a nonintegral ratio if both pinion and gear are made of hardened steel.
9. Make sure critical speeds are at least 20% apart from operating speeds or speed multiples and from frequency of tooth mesh.

FIGURE 2.20 Skew bevel gear set. (From Ref. 5.)

The above recommendations are for metallic gears. No. 8 should
be modified to state that if the gear is made of a softer material, an
integral ratio allows the gear to cold-work and conform to the pinion, thereby promoting quiet operation. The inference once again
is that the gear in this case is to be a softer metal than the metal
pinion rather than soft as applied to plastics.
    A number of the recommendations apply also to noise reduction
in plastics gearing. Pitch selection, pressure angle, profile modification, backlash, quality, balance, and speeds should each be given
due consideration. Both spur and helical gear sets will react to
changes in the above areas.
    The above recommendations do not imply that simply replacing a
metallic gear set with a plastics gear set will resolve all noise problems. The engineer must first determine the source of the noise
whether it is generated in the gears or in the associated components. Damping properties of the chosen plastics material may not
be an improvement. Noise may be generated by bearings, couplings,
or other hardware; equivalent quality plastics gearing may not be
economical or possible; and the noise may not be generated by the
tooth meshing action.

Helical vs. Spur Gear Usage

A helical gear tooth of a given normal diametral pitch is stronger
than a corresponding diametral pitch spur gear because the helical
tooth section is heavier in the plane of rotation.
    A disadvantage of the helical gear is that the shaft bearing must
be designed to carry the axial thrust caused by the helix angle of

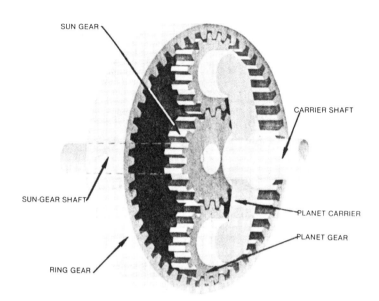

FIGURE 2.21 Planetary gear set. (From Ref. 4.)

the gear teeth. Helical gears are also not as satisfactory as spur gears for critical timing applications, since axial displacement results in additional angular backlash.

Helical gear teeth are more expensive than spur gear teeth. A special master gear for inspection is required for each combination of diametral pitch and helix angle. For spur gears, however, the same master may be used for inspection of all gears of a given diametral pitch.

Of course, the cost of cutting spur and helical gears is not a factor if the gears are to be molded, except in the design of the electrode used in producing the mold. Tooth and angular corrections must be made to the electrode so that the molded parts conform to the exact specified gear when they are molded and ready for the assembly procedure.

As the gear train increases in the number of mesh positions, the complexity of the gear system also increases. Different ratios from mesh to mesh, space necessary for mounting purposes and for shaft bearings, two or more gears on a shaft, and many other factors can tax the ingenuity of the designer. However, after careful consideration of the design requirements and a thorough rating study, plastics may be used to provide gear and component combinations on the same shaft, molded as a unit using the shaft as the bearing, or with two or more gears molded on the same shaft. In some cases little or no lubricant will be required.

A comprehensive reference is recommended, particularly in the subjects of planetary gear design, manufacture, installation, and operation (3). Plastics used in planetary designs again must be applied using the principles discussed. Basic, multistage, and compound planetary gear units can be produced using plastics gearing. Virtually every combination from spur and rack, differential systems to complicated planetary combinations can be produced using plastics throughout or in conjunction with metallic gearing and components.

References

1. D. W. Dudley, Ed., *Gear Handbook*, McGraw-Hill, New York (1962).

2. *"Mechanical Drives Reference Issue,"* Machine Design, Penton/IPC, Cleveland, OH (1983).

3. P. Lynwander, *Gear Drive Systems Design and Application*, Marcel Dekker, New York (1983).

4. J. G. Wills, *Lubrication Fundamentals*, Marcel Dekker, New York (1980).

5. P. M. Dean, Jr., *Gear Manufacture and Performance*, American Society for Metals, Metals Park, OH (1974).

3
Physical Design Criteria

Design recommendations usually follow a procedure such as

1. Determination of the mechanism requirements
2. Preliminary design estimations
3. Short-term material expectations
4. Long-term material property capabilities
5. Stress levels, part strength, and wear resistance
6. Comparison of expected stress levels with the allowable stress levels
7. Design review for molding capability, physical design enhancement, assembly considerations, fabrication and post-fabrication procedures, and cost estimation
8. Sample or prototype application testing
9. Design finalizations, specifications, print review, testing of final assembly with parts from production tooling, approvals, and release for production

Review with the material supplier and the production processor of the plastic part will allow recommendations for change at initial design stages that should eliminate the need for radical design changes further along in the engineering development of the product.

One should not neglect the physical design criteria recommended for the materials chosen for the application. Some of the criteria are subtle but are based on experience and significant test programs conducted by materials suppliers. Assuming that data developed for one material apply to all plastics materials can be disastrous. The following are general recommendations that point out areas where design care is important during material selection.

In comparison with metal gears, plastic gears have a limited resistance to elevated temperatures, a high coefficient of linear thermal
expansion, and a lower load-carrying capacity. Because of these
properties, design consideration is important in relation to shrinkage, warpage, and ability to dissipate heat buildup in both ambient
and operational conditions. Therefore the physical design criteria
are of utmost interest during the initial phases of the design process.

In plastic gear production, molten material is injected into a mold,
or parts are machined from basic stock shapes. Machining of plastic
gears must abide by certain guidelines for heat dissipation during
the machining operation as well as during operation in the gearing
application. Other manufacturing methods are casting, forging,
stamping, and sintering after pressing. AGMA notes that part wall
thickness usually limits the size of injected molded gearing, and the
size of extruded or cast shapes and parts limits the size of machined
gearing. Also, not all plastics materials can be used for the above
manufacturing processes. Consultation between designer and manufacturer will usually provide considerable time savings and mutually
acceptable parts (24).

Shrinkage

Injection molded gears must be easily molded without adverse voids,
parting lines, flash, sprue or ejector pin marks but also must be
easily ejected from the mold. These items must be considered with
the material supplier's data and the molder's cooperation. Material
shrinkage in inch/inch must be known and applied to the part design. Shrinkage rates are available with other necessary data in
molders' and material suppliers' design literature and such publications as *Guide to Plastics Property and Specification Charts* (1),
Modern Plastics Encyclopedia (2), the yearly material reference issue of *Machine Design* (3), *Materials Engineering* (4), *Encyclopedia/
Handbook of Materials, Parts, and Finishes* (5), *The International
Plastics Selector* (6), and materials supplier design handbooks and
manuals (7,8,9). Much data are included in tables but the designer
should be sure the data is the latest available information.

Because shrinkage is a prime characteristic of molded plastic gears,
allowance for this phenomenon will prevent the unexpected failure of
a part. Ribs, walls, radii, tooth size, hub length, hub thickness,
bores, recesses, and subsequent machining or finishing operations
must be considered. Although design alone will not usually eliminate
all warpage or compensate for all shrinkage, employing standard design principles, molders' design recommendations, adjustments made
by the molders, and proper material selection, will insure reasonable

Shrinkage   25

TABLE 3.1 Some Areas of Development and Manufacturing

Concern

Design engineer	Mold producer	Molder	Material supplier
Gear configuration	Material direction flow	Injection speeds, pressures, gate location	Quality
Type of gear and tooth	Removal of flow restrictions		Consistent compounding
Pitch or module	Sprue, runner, flash, ejector pin location	Number of gates	Quality controlled processing
Tip and root modifications			
Profile	Knit line recommendations	Machine finishing, flash removal recommendations, if required, etc.	
Tooth thickness			
Rim, web, hub depth and thickness	Elimination of flow shrinkage differential		
Taper and fillets	Stress riser elimination		
Cross-section uniformity	Contour blending		
Corner radii	Runner balance		
Inserts, if used	Materials Filled or non-filled		
Radii at insert interfaces			
Dimensions and tolerances	Glass or mineral-reinforced, etc.		
Material selection			
Fastening method lubrication, etc.			

plastic parts. Table 3.1 demonstrates that final production of the plastic gear becomes a joint project involving the designer, material supplier, mold designer, and molder. Utilization of good design practice in every phase allows key adjustments as necessary. The design

TABLE 3.2 Warpage Reduction Practices

Mold producer or molder

Adjust injection speed and pressures
Coring, hollow sections in nonfunctional areas
Equal injection force and uniform cooling
Consider size, location of gates, and runner balance

Designer

Use standard design principles
Selection of proper material
Decrease differential flow
Eliminate or decrease nonuniform thickness
Use uniform wall thickness and rib thickness; hold any variation to
 no more than 50% from nominal (60% for glass-reinforced materials);
 blend 2 walls over distance 3X wall thickness
Design with soft contours, large flowing radii, smoothly blending
 planar sections
Eliminate flow interruptions and obstructions
Careful use and location of ribs and flanges: height no more than
 3X wall thickness; tall ribs require 1/4 to 1° taper for mold ejec-
 tion
Give close attention to glass-filled materials and relationship of flow-
 direction to transverse-direction shrinkage (Figure 3.2)
Consider mineral-reinforced or crystalline nylons (low shrinkage,
 reduced warpage, eliminate guess-work in design and processing,
 improved dimensional stability, retention of as-molded properties,
 2 to 3X stiffness improvement)

Source: Ref. 10.

engineer can ensure a good design by continued consultation during
the design phase.

Warpage

Print specifications for a gear are those required for the completely
finished part ready for assembly. Therefore it is necessary that
the moldmaker and molder allow for warpage and shrinkage during

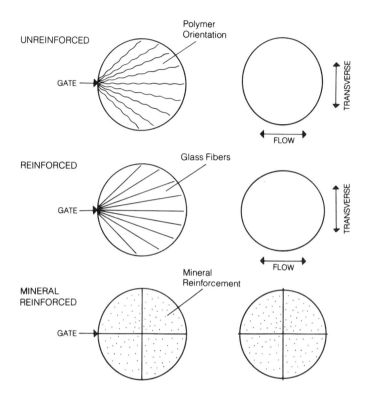

FIGURE 3.1 Warpage/material relationship. Unreinforced: little ani-
sotropic effect. Reinforced: longitudinal shrinkage approximately
50% of transverse shrinkage. Mineral reinforced: uniform shrinkage.
(From Ref. 10.)

gear manufacture. Warpage results from unequal volumetric shrinkage of material as the part cools from a melt to a solid state. A concise discussion of warpage (10) provides key points that are summarized in Table 3.2 and Figure 3.1. These are grouped according
to the person who can best deal with the problem — the design engineer, mold producer, or molder. The consideration of alternate
materials as a warpage or shrinkage solution is suggested. Materials offering low warpage characteristics are listed as follows:

1. Engineering thermoplastic: a mineral-reinforced resin with fine
    basic nodular particulates that provide less orientation, resulting
    in more uniform shrinkage and reduced residual stress.

2. Acetal resin: glass filled; offers very high stiffness and low creep at elevated temperatures.
3. Thermoplastic polyester resin with various percentages of mica and glass by weight.
4. Forty percent mica and glass reinforcement.
5. Materials are offered in both unreinforced and reinforced grades for improved dimensional stability and retention of as-molded properties with improved stiffness of their counterparts at 50% relative humidity.

Advantages are said to be thinner walls than when using 6 or 66 nylons and improved resistance to heat and chemicals. However, consideration of any new material must include the determination of its adequacy and reliability. If a thorough test is not conducted, the best approach may be to use the most widely accepted gear materials and resolve the basic shrinkage and warpage problems.

It is gratifying to see the development of new materials aimed at reducing or eliminating the classical weaknesses of engineering plastics. However, a word of caution is in order about (11) plastics material development and new processes. Experimental-grade plastics can lead to problems because of the following:

1. Small product market leading to discontinuance of material production
2. Poorer mechanical properties
3. Business difficulties such as company bankruptcy, strikes, slowdowns, and energy difficulties; material performance; and processing unknowns and inconsistencies, both material and part production

Alternate processing methods should also be known and considered as possibilities but used with caution (11). Throne suggests that most radically new developments do not work much better than known developments, since they have not had exposure to many conditions and applications. New materials and processes cannot be ignored, however; they must be judiciously considered before development and manufacturing time and costs are expended on a project.

Shrinkage characteristics are well known for the most commonly used plastics for gearing. Various testing programs conducted by materials suppliers have in some cases determined slightly different recommendations for their material (7). A common illustration shown in Figure 3.2 specifies a web thickness of twice the gear-tooth thickness at the pitch circle (12). This agrees with Celcon acetal copolymer (7) whereas DuPont recommends a factor of 1.5t for acetal (8) and nylon (9). Celanese reasons that web thickness can be less than the gear teeth face width without affecting the strength of the gear because

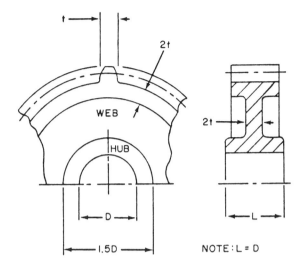

FIGURE 3.2 Suggested gear proportions. (From Ref. 12.)

operating stresses are lower in the web than in the teeth. Added advantages are the reduced cycle times and lessened probability of voids.

The hubless and rimless gear illustrated in Figure 3.3 is recommended for thin gears and has been used successfully on medium-sized gears as well. The probability of warpage due to differential shrinkage is eliminated with this type of configuration.

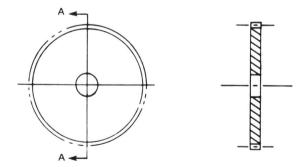

FIGURE 3.3 Hubless and rimless gear. (From Ref. 22.)

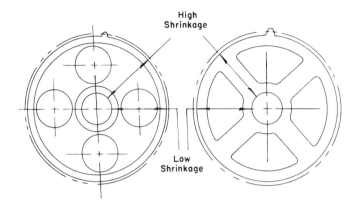

FIGURE 3.4 Holes and ribs in molded gears. (From Ref. 22.)

Reinforced plastics present warpage and shrinkage problems, yet are viable gearing materials. A series of studies presented by LNP Corporation (13) covers such subjects as weld-line integrity (14), fatigue endurance and creep characteristics (15), thermal and environmental resistance (16), effects of fiber orientation (17), impact testing (18), and prediction of shrinkage and warpage (19). A series of processing tips are also given in the "Red Book" (13) for the molder. Richardson (20) and Nielsen (21) provide valuable physical, thermal, and mechanical property data.

Configuration Aspects

Holes and ribs in molded gears, when used for lightening the part or for achieving low inertia in large gears, also present problems of an out-of-round condition typical of parts with nonuniform thicknesses (Figure 3.4). Thin sections in nonuniform gears shrink less than thick sections that cause residual stresses. The compressive load will cause cracks to develop after running the gear at high-tooth loads for several hours. (See Figure 3.5.) In this example the designer should reconsider the necessity for holes or ribs if the out-of-round condition will be detrimental to the gear operation. The characteristic low inertia of plastic gears may not necessitate lightening holes or ribs typical of metallic gears.

    Another important aspect is the method of attaching the gears in the system. The gear and shaft may be molded as a single unit and

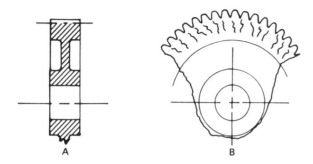

FIGURE 3.5 Failure at high loading on thin web gear. (From Ref. 22.)

the shaft ends seated as journal bearings in the gear box case or mounting side frames. Considerations in this situation are shrinkage, shaft diameter, and length. The length will be subject to shrinkage and the diameter must be such that warpage, after molding or deflection during operation, will not occur or be detrimental to the gearing. Also the gear and shaft material must be compatible with the gear box or side frames and be evaluated as a bearing element.

Inserts

Although there are many advantages and disadvantages of inserts, their prime use in gearing is to hold the plastic gear on a shaft and possibly provide a means of reducing heat buildup generated near the tooth contact surface during operation. Figures 3.6 through 3.8 illustrate stamped, die-cast, and screw machine inserts. The stamped insert is ideal for thin, large diameter gears for reasons of economy, dimensional stability, and rigidity to the plastic gear. Load-carrying capacity of the gear can be increased if a die cast insert is used by using a wider face width plastic gear. A die-cast gear has a thin layer of plastic molded over the gear teeth. Excellent dimensional stability, high load-carrying capacity, and lower cost when compared to a metal cut gear are all advantages. The screw machine insert is more costly than other inserts but provides good concentricity and stability against axial movement and slippage of the molded portion on the insert outer rim. With all inserts, the molded gear teeth must be designed and toleranced so that flash is kept to a minimum. Allowable variations of insert thickness must be closely adhered to, as shown in Figure 3.7. Also the molded teeth must be designed in the mold in a manner that will not cause increased stress as the plastic

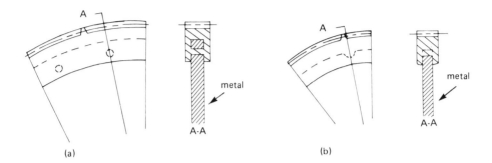

FIGURE 3.6 Stamped metal inserts. (From Ref. 22.)

shrinks around the insert during cooling. Rounded corners, generous radii, and thickness considerations are essential to avoid residual
stresses.

Screws

Fastening gears to shafts can be accomplished by the use of set screws,
but only if disassembly is quite limited. This low-cost method can
easily result in stripped threads, loosening screws, or physical dam-
age to the part by causing stress raisers. Design of the hub must
provide adequate material for sufficient thread engagement as in metal
parts. If frequent disassembly is necessary, a threaded metal insert
can be molded or pressed into the plastic gear.

Keys

Another successful method is the use of round keys when disassembly
is necessary. Moderate cost and ease of disassembly are advantages,
and good torque capacity is possible. For high-load applications, mul-
tiple keys are recommended. Care must be taken that sharp corners
and small inside radii are avoided.

Interference Fits

If a shaft is to be pressed into a plastic gear, knurling provides a low-
cost and frequently used means of preventing the gear from loosening

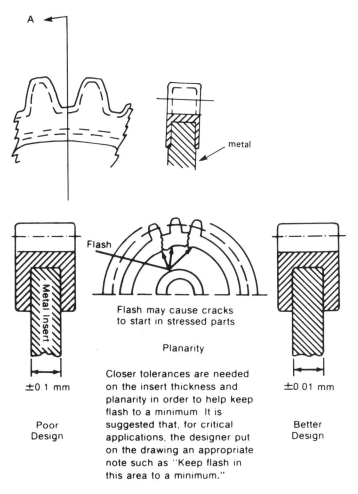

Flash may cause cracks
to start in stressed parts

Planarity

±0.1 mm   Closer tolerances are needed   ±0.01 mm
           on the insert thickness and
           planarity in order to help keep
Poor   flash to a minimum. It is   Better
Design   suggested that, for critical   Design
           applications, the designer put
           on the drawing an appropriate
           note such as "Keep flash in
           this area to a minimum."

FIGURE 3.7 Die-cast insert. (From Ref. 22 and Ref. 9.)

on the shaft in a moderate torque application. Disassembly is never
recommended.

Excellent torque capacity, precise alignment, and good concentricity can be obtained at modest cost by using a splined shaft and
splined teeth molded in the gear hub. All spline corners, edges, and
radii should be rounded according to plastic practices. Standard involute profile and full depth or stub teeth can be used to good advantage and are common when disassembly is frequent.

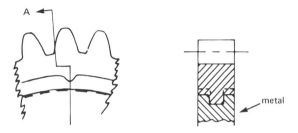

FIGURE 3.8 Screw machine insert. (From Ref. 22.)

Interference fit applications are also common when torque requirements are low. The process is simple, and only the shaft and gear
are needed for this rapid and inexpensive technique. Similar or dis-
similar materials can be used, but limitations may be dictated by the
coefficient of thermal expansion of the two materials. Good design
practice dictates a thorough review of the components and their reactions at time of assembly and effect of time on the joint strength.
    In low torque applications, plastic gears can be pressed onto metal
shafts. Often 50 to 75% of the maximum interference (I) is used as
provided by the following equation when using yield strength of the
plastic material.

$$ I = \frac{S_y \; D_s \left[\left(\dfrac{1+(a/b)^2}{1-(a/b)^2} \right) + \mu \right]}{E \; \left(\dfrac{1+(a/b)^2}{1-(a/b)^2} \right)} \tag{3.1}$$

where

S_y = yield strength of plastic gear material
D_s = metal shaft diameter
E = modulus of elasticity of plastic
a/b = ratio of shaft diameter to hub outside diameter
μ = Poisson's ratio of plastic [some values at 73°F and 50% relative
    humidity are shown in Table 3.3 (12)]

    The procedure can be used for shaft and hub of like plastic ma-
terials, unlike plastic materials, plastic shaft and metal hub, or metal
shaft and plastic hub. Using design stress and safety factors, one
can calculate maximum allowable interference as follows.

TABLE 3.3 Poisson's Ratio μ for Some
Unfilled Thermoplastics

Polymer	μ
Acetal	0.35
Nylon 6/6	0.39
Modified PPO	0.38
Polycarbonate	0.36
Polystyrene	0.33
PVC	0.38
TFE (Tetrafluorethylene)	0.46
FEP (Fluorinated Ethylene Propylene)	0.48

$$I = \frac{S_d D_s}{W}\left[\frac{W + \mu_h}{E_h} + \frac{1 - \mu_s}{E_s}\right] \tag{3.2}$$

and

$$W = \frac{1 + \left(\dfrac{D_s}{D_h}\right)^2}{1 - \left(\dfrac{D_s}{D_h}\right)^2} \tag{3.3}$$

where

I = diametral interference
S_d = design stress = S_t = yield strength at given environmental
conditions divided by a suitable safety factor
S_t = tensile yield strength (Figure 3.9)
D_h = outside diameter of hub
D_s = diameter of shaft
E_h = tensile modulus of elasticity of hub
E_s = modulus of elasticity of shaft

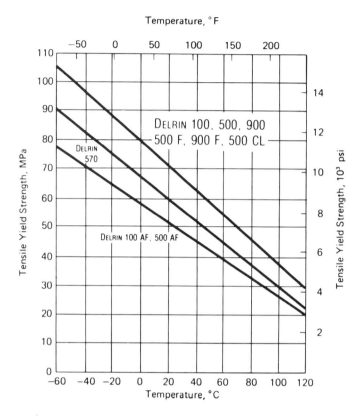

FIGURE 3.9 Tensile yield strength of Delrin vs. temperature (cross-head speed of 5.1 mm/min). (From Ref. 8.)

μ_h = Poisson's ratio of hub material
μ_s = Poisson's ratio of shaft material
W = geometry factor

Although Figures 3.9 and 3.10 were taken from Ref. (8) and Ref. (9) for Delrin® and Zytel®, they illustrate the general case of a plastic material press fit on another plastic material. When similar plastic materials are used for both shaft and gear, Sd is modified by the safety factor and the equation is simplified to

$$I = \frac{Sd \ Ds}{W} \left(\frac{W+1}{Eh} \right) \qquad (3.4)$$

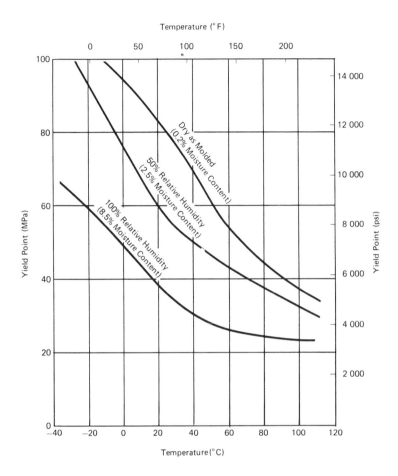

FIGURE 3.10 Yield point of Zytel 101 vs. temperature and moisture
content. (From Ref. 9.)

Metal shafts pressed into plastic hubs present the following when the
modulus of the metal shaft, E is greater than 5×10^6 psi. The last
term in the general case equation becomes negligible and the equation
simplifies to

$$I = \frac{Sd\ Ds}{W} \left(\frac{W+uh}{Eh} \right)$$

(3.5)

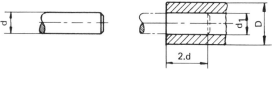

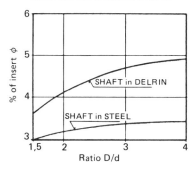

FIGURE 3.11 Maximum interference limits. (From Ref. 8.)

Again, Sd is modified by the factor of safety. In the above cases, 1.35 to 2.0 are typical safety factors. The curves in Figure 3.11 represent maximum allowable interference based on yield point and elastic modulus data at room temperature for press-fitted shafts and hubs of Delrin. Figure 3.12 gives theoretical interference limits for press-fitted Zytel.

In the case of a metal hub and plastic shaft, the general case equation becomes less reliable due to low shaft strength and compressive failure. A safety factor of 1.5 to 3.0 is usually recommended with the following equation

$$I = \frac{Sd\ Ds}{W}\left(\frac{1-\mu s}{Ec}\right) \qquad\qquad (3.6)$$

where Ec is the compressive modulus of the plastic material.

To further evaluate the joint, the force to press the parts together (F) in pounds should be calculated and the resultant joint pressure (P) and torsional strength (T) determined.

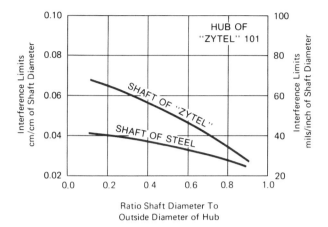

FIGURE 3.12 Theoretical interference limits for press fitting based on yield point and elastic modulus at room temperature and average moisture conditions. (From Ref. 9.)

Approximate equations are:

$$F = \pi f P D_s L \qquad (3.7)$$

$$P = \frac{S_d}{W} \qquad (3.8)$$

$$T = F\left(\frac{D_s}{2}\right) \qquad (3.9)$$

$$D_t - D_o = \alpha \, (t - t_o) D_o \qquad (3.10)$$

where

α = coefficient of linear thermal expansion
f = coefficient of friction
D_o = diameter at initial temperature, t_o
D_s = diameter of shaft
D_t = diameter at temperature, t
L = length of press-fitted surfaces
S_d = design stress = yield strength/safety factor
W = geometry factor from the general case calculation

Coefficient of friction is dependent upon many factors and varies from application to application. It is recommended that tests determine the correct value in cases where greater accuracy is required.

Ease of assembly can be accomplished by heating the external part and/or cooling the internal part to reduce interference among parts. Dimensional changes can be calculated using the coefficient of thermal expansion for the material. If publications and general material property tables do not list values of coefficient of friction, they must be determined by tests or reference to materials supplier data. It is always best to consult the materials supplier if published coefficients are not available or test criteria questionable, or if other test criteria will better suit the application. Tables 3.4 to 3.6 provide coefficients of friction for Delrin and Zytel (8,9,22). Similar table values have been used successfully for reinforced and lubricated thermoplastics (23). The unmodified material data agree well with other published data.

Serious consideration should be given a press-fitted assembly if the effect of time on the joint strength is crucial. A significant reduction of joint pressure and holding power can occur even after one year when Delrin is used. If the expected life of a product is significantly longer than a year, testing and environmental conditions are of major significance. [See Figures 3.13-3.15 (8).] The usual recommendation is to design to the maximum allowable interference due

TABLE 3.4 Coefficient of Friction[a]

	Static	Dynamic
Delrin on steel		
Delrin 100, 500, 900	0.20	0.35
Delrin 500F, 900F	-	0.20
Delrin 500 CL	0.10	0.20
Delrin AF	0.08	0.14
Delrin on Delrin		
Delrin 500/Delrin 500	0.30	0.40
Delrin on Zytel		
Delrin 500/Zytel 101	0.10	0.20

[a]Thrust washer test, non-lubricated, 23°C
(73°F); P, 2.1 MPa (300 psi); V, 3 m/min
(10 fpm).

TABLE 3.5 Range of Coefficients of Friction
on Zytel 101

Zytel on Zytel[a]
No lubricant	Static	Dynamic
Maximum	0.46	0.19
Minimum	0.36	0.11

Zytel on Delrin[a]
No lubricant	Static	Dynamic
Maximum	0.20	0.11
Minimum	0.13	0.08

Zytel on steel
No lubricant	Static	Dynamic
Maximum	0.74	0.43
Minimum	0.31	0.17

Normal pressure: 0.14 MPa (20 psi)
Sliding speed: 0.48 m/s (95 ft/min)
Temperature: 23°C (73°F)
Test method: Thrust washer

(Zytel at 2.5% moisture)

[a]Note: Low thermal conductivity of plastic on
plastic unlubricated pairs reduces PV limit.

TABLE 3.6 Coefficient of Friction of Zytel 101[a]

Lubricant	Other surface	Load MPa	psi	Coefficient of friction
Dry	Zytel	7.2	1050	0.04 to 0.13
Water	Zytel	7.2	1050	0.08 to 0.14
Oil	Zytel	7.2	1050	0.07 to 0.08
Water	Steel	7.2	1050	0.3 to 0.5
Oil	Steel	10.7	1550	0.02 to 0.11
Water	Brass	7.2	1050	0.3 to 0.5
Oil	Brass	10.7	1550	0.08 to 0.14

[a]Battelle Memorial Institute; Neely, or boundary film,
testing machine; surface speed = .8 m/s (156 ft/min)
Source: Ref. 9.

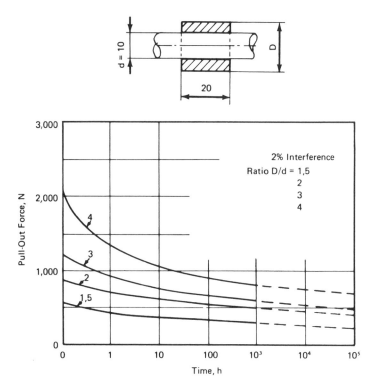

FIGURE 3.13 Time vs. joint strength — 2% interference. (From Ref.
8.)

to necessary manufacturing tolerances, select stress factors from materials property tables, and use a safety factor of 1.5 to 3.0 depending on the application.

In the case of Zytel, smooth-surfaced joint pressure and holding
power are reduced due to creep or stress relaxation. Time effects
may usually be neglected for knurled or grooved parts since creep
or gradual flow counteracts stress relaxation effects.

Mathematical procedures have been used as estimates and predic-
tions for other engineering plastics materials with good success.
Other methods, sometimes novel approaches, have been used to resolve the problems discussed. Usually the plastics industry repre-
sentatives are familiar with them and are available to offer valuable
advice.

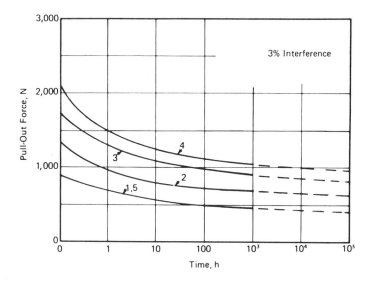

FIGURE 3.14 Time vs. joint strength — 3% interference. (From Ref. 8.)

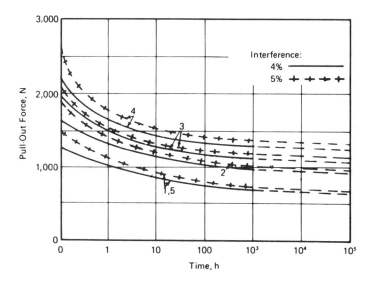

FIGURE 3.15 Time vs. joint strength — 4 and 5% interference. (From Ref. 8.)

There are several mechanical devices on the market that are available for fastening gears on shafts. Before they are used in plastic gearing applications, the candidate designs must be carefully assessed to insure suitable assembly, torque loading, stress levels, and economic feasibility.

References

1. Modern Plastics Encyclopedia, *Guide to Plastics Property and Specification Charts*, McGraw-Hill, New York (1972).

2. *Modern Plastics Encyclopedia*, McGraw-Hill, New York (annual).

3. *Machine Design*, Penton/IPC, Cleveland, OH (weekly).

4. *Materials Engineering*, Reinhold, Stamford, CT (monthly).

5. H. R. Clauser, Ed., *Encyclopedia/Handbook of Materials, Parts, and Finishes*, Technomic, Westport, CT (1977).

6. *Desk-Top Data Bank*, The International Plastics Selector, Inc., San Diego, CA (1983).

7. *Design and Production of Gears in Celcon Acetal Copolymer*, Celanese Plastics Company, Newark, NJ (1969).

8. *DuPont Delrin Acetal Resin Design Handbook*, E. I. DuPont DeNemours and Co., Wilmington, DE (1981).

9. *DuPont Zytel Nylon Resin Design Handbook*, Bulletin E-44971, E. I. DuPont DeNemours and Co., Wilmington, DE.

10. E. W. Lonsbary, "Warpage: How To Deal With It," *Engineering Design with DuPont Plastics*, E. I. DuPont DeNemours and Co., Wilmington, DE (Fall 1982).

11. J. L. Throne, *Plastics Process Engineering*, Marcel Dekker, New York (1979).

12. *Plastics Design Guide*, Corporate Practice (C-P 5-2300-007), IBM Corporation, Armonk, NY (1980).

13. *Fortified Polymers, LNP Engineering Plastics*, Red Book, LNP Corporation, Malvern, PA (1981).

14. P. J. Cloud, F. McDowell, "Reinforced Thermoplastics: Understanding Weld-line Integrity," Reprinted from *Plastics Technology* (August 1976).

15. G. B. Newby, J. E. Theberge, "Long Term Behavior of Reinforced Thermoplastics," Reprinted from *Machine Design*, Penton/IPC, Cleveland, OH (March 1984).

16. P. J. Cloud, J. E. Theberge, "Thermal and Environmental Resistance of Glass-Reinforced Thermoplastics," Adapted from *Machine Design* issues, February 1971–September 1971.

17. *The Effects of Fiber Orientation on Physical Properties*, Reprinted from Plastics Design Forum (September/October 1979).

18. J. E. Theberge, "Impact Testing of Glass-Fortified Thermoplastics," Based on *Modern Plastics* article (July 1969).

19. P. J. Cloud, M. A. Wolverton, "Predict Shrinkage and Warpage of Reinforced and Filled Thermoplastics," Reprinted from *Modern Plastics* (November 1978).

20. E. Miller, Ed., *Plastics Products Design Handbook, Part A, Materials and Components*, Marcel Dekker, NY (1981).

21. L. E. Nielsen, *Mechanical Properties of Polymers and Composites*, Marcel Dekker, NY (1974).

22. *Gears of "Delrin" Acetal Resins and "Zytel" Nylon Resins*, E. I. DuPont DeNemours and Co., Wilmington, DE (1971).

23. *LNP Internally Lubricated Reinforced Thermoplastics*, LNP Corporation, Malvern, PA (1982).

24. *Information Sheet Plastics Gearing — Molded, Machined, and Other Methods* (AGMA 141.01), American Gear Manufacturers Association, Alexandria, VA (1984).

4

Materials and Material Selection

Presented here is general information on plastics materials that are used successfully in various gearing applications. An important note of caution concerns use of the data presented in the tables under their generic names. The following will influence material selection:

1. All the conditions under which tests were made in development of material data are not always available, and the data related to a specific application is usually uncertain.
2. Tooling, manufacturing procedures and methods, equipment used, and variables in each operation are difficult, if not impossible, to predict.
3. Control of material usage and applications resides with the customer rather than the material supplier.
4. Application of the final part is the responsibility of the user.
5. Adequacy of the material in the final application can only be determined by testing in the final assembly. Testing of cut plastics gears will not suffice if molded gears are to be used in final production.

The user should consider the published data cautiously and use prototype testing before final design acceptance is made. Disclaimers by material suppliers are common and reasonable.

The intent here is not to provide all of the technical and chemical engineering details on producing plastics materials. Only a minimum of such information is presented here and in the plastics glossary in Appendix C. Certain terms and concepts must be understood, however, in order to have an appreciation for why some materials perform better than others in the gearing application.

Plastics are manufactured from low-molecular-weight compounds (monomers) that are reacted together to form high-molecular-weight chainlike compounds with a regular repeating structure (mer) that is a fraction of or the whole of a monomer compound. Thermoplastics are chemically bound by strong covalent bonds only within independent long-chain molecules. These long molecules are bound together by weaker intermolecular forces. Thermosetting polymers are three-dimensional networks bound throughout by strong covalent forces, the entire specimen may be considered one giant molecule (1). Molecular weight or molecular size is therefore not a meaningful quantity for thermosets. Mechanical properties of a thermoplastic, however, are determined by a primary factor, average molecular weight. As molecular weight increases, the number of bonds per molecule also increases and the tensile strength, impact strength, fatigue strength, and modulus of elasticity increase. Estabrook (2) discusses mechanical properties of polymers in more detail.

Generally, thermosets are plastic materials that have better dimensional stability, heat resistance, chemical resistance, and electrical properties than do thermoplastics. Most thermosets are used principally in filled and/or reinforced forms, to increase dimensional stability or other properties, or for economy. Most formulations require heat and/or pressure for curing. One thermoset recommended as a structural, mechanical, and wear application candidate is phenolic, a low-cost material with good balance of mechanical, electrical, and thermal properties but limited in color to black and brown. Depending on whether the gear application requires mainly structural or mechanical concerns as opposed to wear, the materials listed in Table 4.1 should be considered.

Thermoplastics generally offer higher impact strength, easier processing, and better adaptability to complex designs than do thermosets.

For gear design, the best materials have a high allowable level of flexural stress, stiffness, and shear strength to combat the tendency to tooth bending and fracture at the tooth root. For unlubricated gears, consider surface stress rather than bending stress. Select materials with a high compressive stress and low coefficient of friction.

Typical Gearing Materials

Certain benefits can be had from various materials. The following can be used for material selection to meet gear application objectives (2,3).

Acetal: very strong, stiff engineering plastic with exceptional dimensional stability due to low moisture absorption, resistance to creep and vibration fatigue; low coefficient of friction; high

TABLE 4.1 Gearing Materials

Structural/mechanical	Wear applications
Acetal	Acetal
Nylon	Nylon
Polyester (thermoplastic)	Polyester (thermoplastic)
Polyimide	Polyimide
Polycarbonate	Polyurethane
Polyphenylene sulfide	UHMW polyethylene

resistance to abrasion and chemicals; retains most properties when immersed in hot water; low tendency to stress-crack.

Nylon: family of engineering resins having outstanding toughness and wear resistance, low coefficient of friction, and excellent electrical properties and chemical resistance. Resins are hygroscopic; dimensional stability is poorer than that of most other engineering plastics.

Phenolic: stability and strength when reinforced with long-fiber fillers.

Polycarbonate: highest impact resistance of any rigid, transparent plastic; excellent outdoor stability and resistance to creep under load; fair chemical resistance; some aromatic solvents cause stress cracking; used with addition of glass fiber reinforcement and/or polytetrafluoroethylene (PTFE) lubricant.

Polyester: excellent dimensional stability, electrical properties, toughness, and chemical resistance, except to strong acids or bases; notch sensitive; not suitable for outdoor use or for service in hot water; also available in thermosetting formulations. Polybutylene terephthalate (PBT) and polyethylene terephthalate (PET) are alternatives to nylon and acetal molding gearing material.

Polyester elastomer: sound dampening; resistance to flex-fatigue and impact.

Polyethylene: wide variety of grades: low-, medium-, and high-density formulations. Low-density types are flexible and tough. Medium- and high-density types are stronger, harder, and more rigid; all are lightweight, easy to process, low-cost materials; poor dimensional stability and heat resistance; excellent chemical resistance and electrical properties. Also available in ultrahigh-molecular weight and in linear low-density grades.

Polyimide: outstanding resistance to heat (500°F continuous, 900°F intermittent) and to heat aging. High-impact strength and wear resistance; low coefficient of thermal expansion; excellent electrical properties; difficult to process by conventional methods; high cost.

Polyphenylene sulfide: outstanding chemical and heat resistance (450°F continuous); excellent low-temperature strength; inert to most chemicals over a wide temperature range; inherently flame retardant; requires high processing temperature.

Polyurethane: tough, extremely abrasive, and impact-resistant material; good electrical properties and chemical resistance; ultraviolet exposure produces brittleness, lower properties, and yellowing; also made in thermoset formulations.

Styrene-acrylonitrile: low-cost for lightly loaded applications needing accuracy and stability.

If the most widely used gear materials will do the job, these should be selected over the lesser known materials. A vast amount of knowledge and experience have been accumulated in plastics gearing. In the interest of time and economy there is usually a well-known material that will adequately suit the user's purpose. The most widely used plastic for gears, molded, machined, or cast, is nylon, which is available in many basic polymer forms. These various forms of nylon are compounded into many different molding materials with several kinds of reinforcements, such as glass fibers, milled glass, glass beads, and carbon fibers; fillers such as minerals and glass; and lubricants such as PTFE, MoS_2, silicone, and graphite.

Acetal is the next most widely used plastic gearing material, with other materials ranking in use as shown in the property comparison chart, Table 4.2. Nevertheless, it is the application and the expected life of the product that ultimately determines the material that should be used. Table 4.3 illustrates how nylon can be modified to meet the necessary requirements of the application.

Material Properties

Many sources (3-15) provide basic material property data. Once a material is chosen, supplier information should be studied in detail. Resin characteristics may vary in melt indices, be inconsistent in percentages of fill materials, have differing mixes of virgin and rework resins, and be susceptible to changes in humidity that may not produce what was intended although the design data used may have been correct. Another design concern involves the complex nature of gearing with regard to the sliding and rolling actions at and near the tooth contact surface. Not all plastics with seemingly ideal material properties work well as gears. Table 4.4 presents materials property data sheets published for 1984-1985 (10).

TABLE 4.2 Property Chart for Basic Polymers for Gearing

	Water absorp. 24 hrs. %	Mold shrinkage in./in.	Tensile strength *yield •break psi	Flexural modulus psi	Izod impact strength notched ft-lb/in.	Deflect. temp. @ 264 psi °F	Coeff. linear thermal expansion 10⁻⁵ °F	Specific gravity
Units	%	in./in.	psi	psi	ft-lb/in.	°F	10⁻⁵ °F	
ASTM	D570	D955	D638	D790	D256	D648	D696	D792
(1) Nylon 6/6	1.5	.015/.030	*11,200	175,000	2.1	220	4.5 varies	1.13/1.15
(2) Nylon 6	1.6	.013/.025	*11,800	395,000	1.1	150	4.6	1.13
(3) Acetal	0.2	.016/.030	*10,000	410,000	1.4/2.3	255	5.8	1.42
(4) Polycarbonate 30% G/F, 15% PTFE	0.06	.0035	•17,500	1,200,000	2	290	1.50	1.55
(5) Polyester (thermoplastic)	0.08	.020	*8000 •12,000	340,000	1.2	130	5.3	1.3
(6) Polyphenylene sulfide 30% G/F 15% PTFE	0.03	.002	*19,000	1,300,000	1.10	500	1.50	1.69
(7) Polyester elastomer	0.3	.012	*3780 •5500	—	—	122	10.00	1.25
(8) Phenolic (molded)	0.45	.007	•7000	340,000	.29	270	3.75	1.42

*These are average values for comparison purposes only.
Source: Ref. 2.

TABLE 4.3 Material Modifications to Improve Application Performance

	Water absorp. (24 hrs)	Mold shrinkage	Tensile strength (#yield; *break)	Flexural modulus	Izod impact strength notched	Deflect. temp. at 264 psi	Coeff. linear thermal expansion	Specific gravity
Units	%	in./in.	psi	psi	ft-lb/in.	°F	10^{-5} °F	
ASTM	D570	D955	D638	D790	D256	D648	D696	D792
Nylon 6/6 Unmodified	1.5	0.015/0.030	11,200	175,000	2.1	220	4.5 varies	1.13/1.15
Nylon 6/6 G/F 30% PTFE 15%	0.5	0.006/0.008	*23,500	1,350,000	1.8	490	2.4	1.49
Nylon 6/6 MoS_2 5%	1.0	0.012	*13,800	450,000	--	230	3.61	1.17

Material								
Nylon 6/6 G/F 40%	0.4	0.002	40,000	3,400,000	1.6	500	0.80	1.34
Nylon 6 Unmodified	1.6	0.013/0.025	#11,800	395,000	1.1	150	4.6	1.13
Nylon 6 G/F 30% PTFE 15%	0.85	0.0035	19,000	1,250,000	1.9	410	2.6	1.47
Nylon 11 Unmodified	0.3	0.0065	#8,250	142,000	0.75	131	5.06	1.04
Nylon 11 Graphite	0.3	NA	*5,700	180,000	NA	149	5.06	1.09
Nylon 12 Unmodified	0.25	0.010/0.015	*7,200 #8,000	213,000	0.90	130	6.11	1.015
Nylon 12 G/F	0.15	0.006/0.010	*9,500 #9,500	498,000	1.5	250	3.89	1.21

G/F = glass fiber; PTFE = polytetrafluoroethylene (lubricant); MoS_2 = molybdenum disulfide (lubricant).

TABLE 4.4 Resins and Compounds

	Properties	ASTM test method	ABS — Extrusion grade	Flame-retarded grades, molding and extrusion — ABS	Flame-retarded grades, molding and extrusion — ABS/PVC	Flame-retarded grades, molding and extrusion — ABS/PC	ABS/PC injection molding and extrusion	Injection molding grades — Heat-resistant	Injection molding grades — Medium-impact
Processing	1. Melting temperature, °C. T_m (crystalline)								
	T_g (amorphous)		88-120	110-125				110-125	105-115
	2. Processing temperature range, °F. (C = compression; T = transfer; I = injection; E = extrusion)		E: 350-450	C: 350-500 I: 380-500		I: 425-500	I: 460-525 E: 450-500	C: 325-500 I: 475-550	C: 325-350 I: 390-525
	3. Molding pressure range, 10^3 p.s.i.			8-25		10-20	10-20	8-25	8-25
	4. Compression ratio		2.5-2.7	1.1-2.0	2.0-2.5	1.1-2.5	1.1-2.0	1.1-2.0	1.1-2.0
	5. Mold (linear) shrinkage, in./in.	D955		0.004-0.008	0.003-0.005	0.005-0.007	0.005-0.008	0.004-0.009	0.004-0.009
Mechanical	6. Tensile strength at break, p.s.i.	D638[b]	2500-8000	5000-8000	5800	6800-9300	7100-7400	5000-7500	5500-7500
	7. Elongation at break, %	D638[b]	20-100	5-25		50	50-65	3-30	5-25
	8. Tensile yield strength, p.s.i.	D638[b]	4300-6400	4000-7400	5800	8400-9000	3500-8500	4300-7000	5000-6500
	9. Compressive strength (rupture or yield), p.s.i.	D695	5200-10,000	6500-7500			11,200-11,300	7200-10,000	1800-12,500
	10. Flexural strength (rupture or yield), p.s.i.	D790	4000-14,000	9000-14,000	9200-9600	12,000-13,600	12,000-13,000	9500-13,000	7100-13,000
	11. Tensile modulus, 10^3 p.s.i.	D638[b]	130-400	300-420	325-330	370-455	350-380	300-360	300-400
	12. Compressive modulus, 10^3 p.s.i.	D695	150-390	130-310		230		190-440	200-450
	13. Flexural modulus, 10^3 p.s.i. 73° F.	D790	130-440	300-410	320-340	350-387	320-370	300-400	310-400
	200° F.	D790							
	250° F.	D790							
	300° F.	D790							
	14. Izod impact, ft.-lb./in. of notch (¼-in. thick specimen)	D256A	1.8-12	3.0-12	6.5-10.5	4.1-10.5	6.4-10.5	2.0-6.5	3.0-6.0
	15. Hardness Rockwell	D785	R75-115	R100-120	R100-106	R117-119	R111-120	R100-115	R107-115
	Shore/Barcol	D2240/D2583							
Thermal	16. Coef. of linear thermal expansion, 10^{-6} in./in./°C.	D696	60-130	65-95	46	67	62-72	60-93	80-100
	17. Deflection temperature under flexural load, °F. 264 p.s.i.	D648	170-220 annealed	195-225 annealed	180	211-220	232-240	220-240 annealed	200-220 annealed
	66 p.s.i.	D648	170-235 annealed	210-245 annealed		225-244	225-250	230-245 annealed	215-225 annealed
	18. Thermal conductivity, 10^{-4} cal.-cm./sec.-cm².-°C.	C177					4.5-8.0		
Physical	19. Specific gravity	D792	1.02-1.06	1.16-1.21	1.20-1.21	1.2-1.23	1.07-1.12	1.05-1.08	1.03-1.06
	20. Water absorption (¼-in. thick specimen), % 24 hr.	D570	0.20-0.45	0.2-0.6		0.24	0.21-0.24	0.20-0.45	0.20-0.45
	Saturation	D570							
	21. Dielectric strength (⅛-in. thick specimen), short time, v./mil	D149	350-500	350-500	500	450	430	350-500	350-500
	SUPPLIERS[a]		Schulman Shuman Borg-Warner; Dow Chem.; Monsanto	Schulman Borg-Warner; Monsanto	Schulman USS Chem.; Borg-Warner; Monsanto	Borg-Warner; Mobay	Borg-Warner; Mobay	Shuman Borg-Warner; Dow Chem.; Monsanto; Montedison	Schulman Shuman Borg-Warner; Dow Chem.; Monsanto; Montedison

[a]See Ref. 10 for suppliers of specialty materials and custom compounds.
[b]Tensile test method varies with material: D 638 is standard for thermoplastics; D 651 for rigid thermosetting plastics; D 412 for elastomeric plastics; D 882 for thin plastics sheeting.
Source: Ref. 10.

TABLE 4.4 (continued)

	ABS (Cont'd)						Acetal		
	Injection molding grades (Cont'd)			EMI shielding (conductive)					
	High-impact	Platable grade	20% glass-reinforced	20% PAN carbon fiber	20% graphite fiber	40% aluminum flake	Homopolymer	Copolymer	Impact-modified homopolymer
1.							175-181	175	175
	100-110	100-110	100-110						
2.	C: 325-350 I: 380-525	C: 325-400 I: 350-500	C: 350-500 I: 350-500	I: 430-500	I: 420-530	I: 450-550	I: 380-470	C: 340-400 I: 360-450	I: 380-430
3.	8-25	8-25	15-30				10-20	10-20	6-12
4.	1.1-2.0	1.1-2.0					3.0-4.5	3.0-4.0	
5.	0.004-0.009	0.005-0.008	0.002	0.0005-0.003	0.001	0.001	0.020-0.025	0.020 (Avg.)	0.012-0.019
6.	4400-6300	6000-6400	11,000	16,000	15,200-15,800	4200	9700		6500-8400
7.	5-70		3	1.0	2.0-2.2	5	13-75	40-75	60-200
8.	2600-5900						9500-12,000	8800	
9.	4500-8000		14,000		16,000-17,000	6500	15,600-18,000 @ 10%	16,000 @ 10%	
10.	5400-11,500	10,500-11,500	14,000-15,500	25,000	23,000	7800	13,600-14,000	13,000	
11.	150-330	330-380	740	2000	1660	370	520	410	
12.	140-300						670	450	
13.	179-375	340-390	650-710	1800	1560	400	380-430	375	200-350
									90-190
14.	6.0-9.3	5.0-5.3	1.2-1.4	1.0	1.3	2.0	1.2-2.3	1.0-1.5	2.1-17
15.	R85-106	R103-109	M85			R107	M92-94	M78	M58-79
16.	95-110	47-53	21		20	40	100	85	110-122
17.	205-215 annealed	204-215 annealed	210	215	216	212	255-260	230	194-212
	210-225 annealed	215-222 annealed	220		240	220	328-338	316	293-336
18.							5.5	5.5	
19.	1.01-1.05	1.06-1.07	1.22	1.14	1.17	1.61	1.42	1.41	1.34-1.39
20.	0.20-0.45				0.15	0.23	0.25-0.40	0.22	
							1.41		
21.	350-500	420-550	460				500 (90 mil)	500 (90 mil)	400-480
	Schulman Shuman Borg-Warner; Dow Chem; Monsanto; Montedison	USS Chem; Borg-Warner; Monsanto	RTP Schulman LNP; Thermofil; Washington Penn; Wilson-Fiberfil	Wilson-Fiberfil	Thermofil	Thermofil	Du Pont	Celanese	Du Pont

(continued)

Materials and Material Selection

TABLE 4.4 (continued)

Materials	Properties	ASTM test method	Acetal (Cont'd) 20% glass-reinforced homopolymer	25% glass coupled copolymer	21% PTFE-filled homo-polymer	5-20% PTFE-filled copolymer	EMI shielding (conductive); 30% pitch carbon fiber	Acrylic Sheet[c] Cast	Cast, flame-retarded
Processing	1. Melting temperature, °C. T_m (crystalline)		181	175	181	175			
	T_g (amorphous)							90-105	90-105
	2. Processing temperature range, °F. (C = compression; T = transfer; I = injection; E = extrusion)		I: 350-480	I: 380-480	I: 370-410	I: 350-410	I: 350-400		
	3. Molding pressure range, 10^3 p.s.i.		10-20	10-20	10-20	10-20			
	4. Compression ratio				3.0-4.0				
	5. Mold (linear) shrinkage, in./in.	D955	0.009-0.012	0.004 (flow) 0.018 (trans.)	0.020-0.025	0.020-0.029	0.003-0.005		
Mechanical	6. Tensile strength at break, p.s.i.	D638[b]	8500	18,500		6300-6600	7500	8000-11,000	8000-12,500
	7. Elongation at break, %	D638[b]	7	3	15-22	6-8	1.5	2-7	4-5
	8. Tensile yield strength, p.s.i.	D638[b]			6900-7600	6400			
	9. Compressive strength (rupture or yield), p.s.i.	D695	18,000 @ 10%	17,000 @ 10%	13,000 @ 10%	11,000-12,600		11,000-19,000	11,000-12,000
	10. Flexural strength (rupture or yield), p.s.i.	D790	15,000	28,000		9300-10,100	12,500	12,000-17,000	12,000-18,000
	11. Tensile modulus, 10^3 p.s.i.	D638[b]	1000	1250		250-280	1350	350-450	350-480
	12. Compressive modulus, 10^3 p.s.i.	D695						390-475	450
	13. Flexural modulus, 10^3 p.s.i. 73° F.	D790	730	1100	340-350	310-360	1100	390-475	350-450
	200° F.	D790							
	250° F.	D790							
	300° F.	D790							
	14. Izod impact, ft.-lb./in. of notch (¼-in. thick specimen)	D256A	0.8	1.8	0.7-1.2	0.5-1.0	0.7	0.3-0.4	0.3-0.4
	15. Hardness Rockwell	D785	M90	M79	M78	M51-61		M80-100	M61-100
	Shore/Barcol	D2240/ D2583							
Thermal	16. Coef. of linear thermal expansion, 10^{-6} in./in./°C.	D696	36-81		75	52-68		50-90	77
	17. Deflection temperature under flexural load, °F. 264 p.s.i.	D648	315	325	212	198-215	320	160-215	155-205
	66 p.s.i.	D648	345	331	329	306-325		165-235	170-200
	18. Thermal conductivity, 10^{-4} cal.-cm./ sec.-cm.^{2}-°C.	C177						4.0-6.0	4.0-6.0
Physical	19. Specific gravity	D792	1.56	1.61	1.54	1.44-1.51	1.53	1.17-1.20	1.21-1.28
	20. Water absorption (⅛-in. thick specimen), % 24 hr.	D570	0.25	0.29	0.20	0.23-0.26	0.26	0.2-0.4	0.3-0.4
	Saturation	D570							
	21. Dielectric strength (⅛-in. thick specimen), short time, v./mil	D149	490	580	400	410		450-550	400-440
	SUPPLIERS[a]		Du Pont; LNP; Thermofil; Wilson-Fiberfil	Celanese; LNP; Thermofil; Wilson-Fiberfil	Du Pont; LNP; Thermofil; Wilson-Fiberfil	Thermofil; Wilson-Fiberfil	Wilson-Fiberfil	Du Pont; Rohm & Haas	Rohm & Haas

[a]See Ref. 10 for suppliers of specialty materials and custom compounds.
[b]Tensile test method varies with material: D638 is standard for thermoplastics; D651 for rigid thermosetting plastics; D412 for elastomeric plastics; D882 for thin plastics sheeting.
[c]The data for cast acrylic sheet should be used as a guide only since the materials produced by the various manufacturers differ considerably.

TABLE 4.4 (continued)

	Acrylic (Cont'd)							Acrylo-nitrile	Allyl
	Sheet^c (Cont'd)			Molding compounds					
	High-impact	Coated	Acrylic-PVC alloy	PMMA	MMA-styrene copolymer	Impact-modified	Heat-resistant	Molding and extrusion	Allyl diglycol carbonate cast sheet
1.	90-100	90-110	105	90-105	100-105	80-100	100-125	135 / 95	Thermoset
2.				C: 300-425 I: 325-500 E: 360-507	C: 300-400 I: 330-500	C: 300-400 I: 400-500	C: 350-450 I: 400-500 E: 360-507	C: 320-345 I: 410 E: 350-400	
3.				5-20	10-30	10-20	6-30	20	
4.				1.6-3.0			1.2-2.0	2	
5.				0.001-0.004(flow) 0.002-0.008(trans.)	0.002-0.006	0.004-0.008	0.003-0.008	0.002-0.005	
6.	6500	10,500	6500	7000-11,000	10,000	5000-9000	9300-10,000	9000	5000-6000
7.	40-140	3	100	2-10	3	20-70	2-10	3-4	
8.									
9.		18,000	8400	10,500-18,000	11,000-15,000	4000-14,000	15,000-17,000	12,000	21,000-23,000
10.	8900	16,000	10,700	10,500-19,000	16,000-19,000	7000-13,000	12,000-18,000	14,000	6000-13,000
11.	265	450	330-335	325-470	430	200-400	350-470	510-580	300
12.		450	330-400	370-460	240-370	240-370			300
13.	300	450	330-400	325-460	260-380	200-380	460-500	500-590	250-330
14.	1.4	0.3-0.4	15	0.3-0.6	0.3	0.8-2.5	0.3-0.4	1.5-4.8	0.2-0.4
15.	M61	M105	R99-105	M68-105	M75	R105-120	M94-105	M72-78	M95-100
16.		40		50-90	60-80	50-80	50-71	66	81-143
17.	172	205	160	165-210	205-212	165-203	190-221	164	140-190
	195	225	177	175-225		180-205	205-240	172	
18.		5.0		4.0-6.0	4.0-5.0	4.0-5.0	4.5	6.2	4.8-5.0
19.				1.17-1.20	1.09	1.11-1.18	1.16-1.19	1.15	1.3-1.4
20.		<0.4	0.06	0.1-0.4	0.15	0.2-0.8	0.2-0.3	0.38	0.2
21.		500	>400	400-500	450	380-500	400-500	220-240	380
	Rohm & Haas	Rohm & Haas	Rohm & Haas	Continental Polymers Cyro Du Pont; Rohm & Haas Richardson	Richardson	Continental Polymers Cyro Du Pont; Rohm & Haas	Continental Polymers Cyro Rohm & Haas	Sohio	PPG

(continued)

Materials and Material Selection

TABLE 4.4 (continued)

| | | Allyl (Cont'd) | | Cellulosic | | | |
| | | DAP molding compounds | | | Cellulose acetate | | Cellulose acetate butyrate |
Properties	ASTM test method	Glass-filled	Mineral-filled	Ethyl cellulose molding compound and sheet	Sheet	Molding compound	Sheet
Processing							
1. Melting temperature, °C. T_m (crystalline)		Thermoset	Thermoset	135	230	230	140
T_g (amorphous)							
2. Processing temperature range, °F. (C = compression; T = transfer; I = injection; E = extrusion)		C: 290-280 I: 300-350	C: 270-380	C: 250-390 I: 350-500		C: 260-420 I: 335-490	
3. Molding pressure range, 10^3 p.s.i.		2000-6000	2500-5000	8-32		8-32	
4. Compression ratio		1.9-10.0	1.2-2.3	1.8-2.4		1.8-2.6	
5. Mold (linear) shrinkage, in./in.	D955	0.0005-0.005	0.002-0.007	0.005-0.009		0.003-0.010	
Mechanical							
6. Tensile strength at break, p.s.i.	D638[b]	6000-11,000	5000-8000	2000-8000	4500-8000	1900-9000	2600-6900
7. Elongation at break, %	D638[b]	3-5	3-5	5-40	20-50	6-70	50-100
8. Tensile yield strength, p.s.i.	D638[b]						
9. Compressive strength (rupture or yield), p.s.i.	D695	25,000-35,000	20,000-32,000			3000-8000	
10. Flexural strength (rupture or yield), p.s.i.	D790	9000-20,000	8500-11,000	4000-12,000	6000-10,000	2000-16,000	4000-9000
11. Tensile modulus, 10^3 p.s.i.	D638[b]	1400-2200	1200-2200				200-250
12. Compressive modulus, 10^3 p.s.i.	D695						
13. Flexural modulus, 10^3 p.s.i. 73°F.	D790	1200-1500	1000-1400			1200-4000	
200°F.	D790						
250°F.	D790						
300°F.	D790						
14. Izod impact, ft.-lb./in. of notch (1/8-in. thick specimen)	D256A	0.4-15.0	0.3-0.8	0.4	2.0-8.5	1.0-7.8	
15. Hardness Rockwell	D785	E80-87	E61	R50-115	R85-120	R34-125	R30-115
Shore/Barcol	D2240/ D2583						
Thermal							
16. Coef. of linear thermal expansion, 10^{-6} in./in./°C.	D696	10-36	10-42	100-200	100-150	80-180	110-170
17. Deflection temperature under flexural load, °F. 264 p.s.i.	D648	330-550+	320-550	115-190		111-195	
66 p.s.i.	D648					120-209	
18. Thermal conductivity, 10^{-4} cal.-cm./ sec.-cm.^{2}.°C.	C177	5.0-15.0	7.0-25	3.8-7.0	4-8	4-8	4-8
Physical							
19. Specific gravity	D792	1.70-1.98	1.65-1.85	1.09-1.17	1.28-1.32	1.22-1.34	1.15-1.22
20. Water absorption (1/8-in. thick specimen), % 24 hr.	D570	0.12-0.35	0.2-0.5	0.8-1.8	2.0-7.0	1.7-6.5	0.9-2.2
Saturation	D570						
21. Dielectric strength (1/8-in. thick specimen), short time, v./mil	D149	400-450	400-450	350-500	250-600	250-600	250-400
SUPPLIERS[a]		Occidental; Cosmic Plastics; Plaskon; Rogers Corp.; U.S. Prolam	Occidental; Cosmic Plastics; Plaskon; Rogers Corp.; U.S. Prolam	Am. Polymers; Dow Chem.; Himont USA	Eastman; Rotuba Plastics	Eastman; Am. Polymers; Rotuba Plastics	Eastman

[a]See Ref. 10 for suppliers of specialty materials and custom compounds.
[b]Tensile test method varies with material: D638 is standard for thermoplastics; D651 for rigid thermosetting plastics; D412 for elastomeric plastics; D882 for thin plastics sheeting.

TABLE 4.4 (continued)

Cellulosic (Cont'd)			Chlorinated PE	Epoxy				
Cellulose acetate butyrate (Cont'd)				Bisphenol molding compounds			Novolak molding compounds	
Molding compound	Cellulose acetate propionate molding compound	Cellulose nitrate	36-42% Cl extrusion grade	Glass fiber-reinforced	Mineral-filled	Low density glass sphere-filled	Mineral- and glass-, filled, encapsulation	Mineral- and glass-filled, high temperature
1. 140	190	125		Thermoset	Thermoset	Thermoset	Thermoset	Thermoset
							145-155	195
2. C: 265-390 I: 335-480	C: 265-400 I: 335-515	C: 185-250	E: 300-400	C: 300-330 T: 280-380	C: 250-330 T: 250-380	C: 250-300 I: 250-300	C: 280-360 I: 290-350 T: 250-380	T: 340-400
3. 8-32	8-32	2-5		1-5	0.1-3	0.1-2	0.25-3.0	0.5-2.5
4. 1.8-2.4	1.8-3.4			3.0-7.0	2.0-3.0	3.0-7.0		1.5-2.5
5. 0.003-0.009	0.003-0.009			0.001-0.008	0.002-0.010	0.006-0.010	0.004-0.008	0.004-0.007
6. 2600-6900	2000-7800	7000-8000	1500-1800	5000-20,000	4000-10,800	2500-4000	5000-12,500	6000-15,500
7. 40-88	29-100	40-45	600-800	4				
8.								
9. 2100-7500	2400-7000	2100-8000		18,000-40,000	18,000-40,000	10,000-15,000	24,000-48,000	30,000-48,000
10. 1800-9300	2900-11,400	9000-11,000		8000-30,000	6000-18,000	5000-7000	10,000-21,800	10,000-21,800
11. 50-200	60-215	190-220		3000			2100	2300-2400
12.					650			660
13. 90-300	120-350			2000-4500	1400-2000	500-750	1400-2400	2300-2400
14. 1.0-10.9	0.5-No break	5-7		0.3-10.0	0.3-0.5	0.15-0.25	0.3-0.5	0.4-0.45
15. R31-116	R10-122	R95-115		M100-112	M100-M112		M115	
			Shore A65				Barcol 70-74	Barcol 78
16. 110-170	110-170	80-120		11-50	20-60		18-43	35
17. 113-202	111-228	140-160		225-500	225-500	200-250	310-446	500
130-227	147-250							
18. 4-8	4-8	5.5		4.0-10.0	4-35	4.0-6.0	8-31	17-24
19. 1.15-1.22	1.17-1.24	1.35-1:40		1.6-2.0	1.6-2.1	0.75-1.0	1.6-2.05	1.85-1.94
20. 0.9-2.2	1.2-2.8	1.0-2.0		0.04-0.20	0.03-0.20	0.2-1.0	0.04-0.29	0.17
							0.15-0.3	
21. 250-400	300-450	300-600		250-400	250-420	380-420	325-450	440-450
Eastman	Eastman	Chem. Development; George, P.D.	Dow Chem.	Fiberite Morton; Hysol; M & T	Fiberite Morton; Hysol; M & T; Plaskon	Fiberite Hysol	Fiberite Morton; Cosmic Plastics; Hysol; M & T; Plaskon; U.S. Prolam	Cosmic Plastics; Plaskon; U.S. Prolam

(continued)

TABLE 4.4 (continued)

Materials	Properties	ASTM test method	Novolak MC (Cont'd) Glass-filled, high strength	Unfilled	Silica-filled	Aluminum-filled	Flexibilized	Cyclo-aliphatic	Fluoro-plastics Polychloro-trifluoro-ethylene
			Epoxy (Cont'd) — Casting resins and compounds						
Processing	1. Melting temperature, °C T_m (crystalline)		Thermoset	Thermoset	Thermoset	Thermoset	Thermoset	Thermoset	220
	T_g (amorphous)								
	2. Processing temperature range, °F (C = compression; T = transfer; I = injection; E = extrusion)		C: 290-330 T: 290-330						C: 460-580 I: 500-600
	3. Molding pressure range, 10^3 p.s.i.		2.5-5.0						15-60
	4. Compression ratio		6-7						2.6
	5. Mold (linear) shrinkage, in./in.	D955	0.0002	0.001-0.010	0.0005-0.008	0.001-0.005	0.001-0.010		0.010-0.015
Mechanical	6. Tensile strength at break, p.s.i.	D638[b]	18,000-27,000	4000-13,000	7000-13,000	7000-12,000	2000-10,000	8000-12,000	4500-6000
	7. Elongation at break, %	D638[b]		3-6	1-3	0.5-3	20-85	2-10	80-250
	8. Tensile yield strength, p.s.i.	D638[b]							5300
	9. Compressive strength (rupture or yield), p.s.i.	D695	30,000-38,000	15,000-25,000	15,000-35,000	15,000-33,000	1000-14,000	15,000-20,000	4600-7400
	10. Flexural strength (rupture or yield), p.s.i.	D790	50,000-70,000	13,000-21,000	8000-14,000	8500-24,000	1000-13,000	10,000-13,000	7400-11,000
	11. Tensile modulus, 10^3 p.s.i.	D638[b]	350				1-350	495	150-300
	12. Compressive modulus, 10^3 p.s.i.	D695							170-200
	13. Flexural modulus, 10^3 p.s.i. 73° F.	D790	2.8-4.2						190-260
	200° F.	D790							
	250° F.	D790							
	300° F.	D790							
	14. Izod impact, ft.-lb./in. of notch (⅛-in. thick specimen)	D256A	25-34	0.2-1.0	0.3-0.45	0.4-1.6	2.3-5.0		2.5-5
	15. Hardness Rockwell	D785		M80-110	M85-120	M55-85			R75-95
	Shore/Barcol	D2240/D2583	Barcol 60-74					Shore D65-89	Shore D75-80
Thermal	16. Coef. of linear thermal expansion, 10^{-6} in./in./°C.	D696		45-65	20-40	5.5	20-100		36-70
	17. Deflection temperature under flexural load, °F 264 p.s.i.	D648		115-550	160-550	190-600	73-250	200-450	
	66 p.s.i.	D648							258
	18. Thermal conductivity, 10^{-4} cal.-cm./ sec.-cm.^{2}.°C.	C177		4.5	10-20	15-25			4.7-5.3
Physical	19. Specific gravity	D792	1.84	1.11-1.40	1.6-2.0	1.4-1.8	0.96-1.35	1.16-1.21	2.08-2.2
	20. Water absorption (⅛-in. thick specimen), % 24 hr.	D570		0.08-0.15	0.04-0.1	0.1-4.0	0.27-0.5		0
	Saturation	D570							
	21. Dielectric strength (⅛-in. thick specimen), short time, v./mil	D149	380-400	300-500	300-550			235-400	500-600
	SUPPLIERS[a]		U.S. Prolam	Ciba-Geigy; Devcon; Dow Chem.; Emerson & Cuming; Epic Resins; Hysol; Isochem; Shell; Thermoset Plastics	Emerson & Cuming; Epic Resins; Hysol; Isochem; Thermoset Plastics	Devcon; Emerson & Cuming; Isochem; Thermoset Plastics	Dow Chem.; Emerson & Cuming; Hysol; Isochem; Thermoset Plastics	Union Carbide	Allied 3M

[a]See Ref. 10 for suppliers of specialty materials and custom compounds.
[b]Tensile test method varies with material: D638 is standard for thermoplastics; D651 for rigid thermosetting plastics; D412 for elastomeric plastics; D882 for thin plastics sheeting.

TABLE 4.4 (continued)

	Fluoroplastics (Cont'd)								
	Polytetrafluoroethylene			**Fluorinated ethylene propylene**		**Polyvinylidene fluoride**		**Modified PE-TFE**	
	Granular	25% glass fiber-reinforced	PFA fluoroplastic	Unfilled	20% milled glass fiber	Injection molding and extrusion	EMI shielding (conductive); 30% PAN carbon fiber	Unfilled	25% glass fiber-reinforced
1.	327	327	310	275	262	156-170		270	270
						-30 to -20			
2.			C: 625-700 I: 700-800	C: 600-750 I: 625-760	I: 600-700	C: 450-500 I: 450-500 E: 450-550	I: 430-500	C: 575-625 I: 570-650	C: 575-625 I: 570-650
3.	2-5	3-8	3-20	5-20	10-20	2-10		2-20	2-20
4.	2.5-4.5		2.0	2.0		3			
5.	0.030-0.060	0.018-0.020	0.040	0.030-0.060	0.006-0.010	0.020-0.030	0.001	0.030-0.040	0.002-0.030
6.	2000-5000	2000-2700	4000-4300	2700-3100	2400	4900-6200	14,000	6500	12,000
7.	200-400	200-300	300	250-330	5	80-300	0.8	100-400	8
8.						6000-6400			
9.	1700			2200		9700-14,000		7100	10,000
10.		2000			4000	8600-9700	19,800	5500	10,700
11.	58-80			50		190-220	2800	120	1200
12.	60								
13.	80	235	120	80-95	250	290-360	2100	200	950
								80	450
								60	310
								20	200
14.	3	2.7	No break	No break	3.2	3-7	1.5	No break	9.0
15.						R77-83		R50	R74
	Shore D50-55	Shore D60-70	Shore D64	Shore D60-65		Shore D75-77		Shore D75	
16.		77-100			22	70-140		59	10-32
17.					150	189-239	318	160	410
	250		166	158				220	510
18.	6.0	8-10	6.0	6.0		2.4-3.0		5.7	
19.	2.14-2.20	2.2-2.3	2.12-2.17	2.12-2.17		1.76-1.78	1.74	1.7	1.8
20.	<0.01		0.03	<0.01	0.01	0.04-0.06	0.12	0.03	0.02
21.	480	320	500	500-600		260-280		400	425
	Allied Du Pont; ICI Americas Am. Hoechst, Montedison	**Allied** Du Pont; ICI Americas Am. Hoechst; LNP; Montedison	**Du Pont**	**Du Pont**	LNP	**Pennwalt;** Kay-Fries; Soltex	Wilson-Fiberfil	**Du Pont**	**Du Pont;** LNP

(continued)

Materials and Material Selection

TABLE 4.4 (continued)

	Properties	ASTM test method	Fluoroplastics (Cont'd) Modified PE-TFE (Cont'd) — EMI shielding (conductive); 20% PAN carbon fiber	PE-CTFE	Furan Asbestos-filled	Ionomer Molding and extrusion	Ionomer Glass- and rubber-modified; molding and extrusion	Melamine formaldehyde Cellulose-filled
Processing	1. Melting temperature, °C. T_m (crystalline)			220-245	Thermoset	81-96	81-96	Thermoset
	T_g (amorphous)							
	2. Processing temperature range, °F. (C = compression; T = transfer; I = injection; E = extrusion)			C: 500 I: 525-575 E: 500-550	C: 275-300	C: 280-350 I: 300-550 E: 300-450	C: 300-400 I: 300-550 E: 350-525	C: 280-370 I: 200-340 T: 300
	3. Molding pressure range, 10³ p.s.i.			5-20	0.1-0.5	2-20	2-20	8-20
	4. Compression ratio					3	3	2.1-3.1
	5. Mold (linear) shrinkage, in./in.	D955	0.001-0.005	0.020-0.025		0.003-0.010	0.002-0.008	0.005-0.015
Mechanical	6. Tensile strength at break, p.s.i.	D638[b]	13,000	6000-7000	3000-4500	2500-5400	3500-7900	5000-13,000
	7. Elongation at break, %	D638[b]	2.0	200-300		300-500	5-100	0.6-1
	8. Tensile yield strength, p.s.i.	D638[b]		4500-4900		1200-2200	1200-2900	
	9. Compressive strength (rupture or yield), p.s.i.	D695			10,000-13,000			33,000-45,000
	10. Flexural strength (rupture or yield), p.s.i.	D790	17,000	7000	600-9000			9000-16,000
	11. Tensile modulus, 10³ p.s.i.	D638[b]	1500	240	1580	20-60		1100-1400
	12. Compressive modulus, 10³ p.s.i.	D695						
	13. Flexural modulus, 10³ p.s.i. 73° F.	D790	1300	240		14-55	8-700	1100
	200° F.	D790						
	250° F.	D790						
	300° F.	D790						
	14. Izod impact, ft.-lb./in. of notch (⅛-in. thick specimen)	D256A	2.8	No break		7-12.6-No break	2.5-No break	0.2-0.4
	15. Hardness Rockwell	D785		R93-95	R110	R53		M115-125
	Shore/Barcol	D2240/ D2583		Shore D75		Shore D54-66	Shore D43-70	
Thermal	16. Coef. of linear thermal expansion, 10⁻⁶ in./in./°C.	D696		80		100-170	50-100	40-45
	17. Deflection temperature 264 p.s.i. under flexural load, °F.	D648	425	170		93-100	111-158	350-390
	66 p.s.i.	D648	435	240		113-125	131-180	
	18. Thermal conductivity, 10⁻⁴ cal.-cm./sec.-cm.²-°C.	C177		3.8		5.7-6.6		6.5-10
Physical	19. Specific gravity	D792	1.72	1.68-1.69	1.75	0.93-0.96	0.95-1.2	1.47-1.52
	20. Water absorption (⅛-in. thick specimen), % 24 hr.	D570	0.10	0.01	0.01-0.2	0.1-0.5	0.1-0.5	0.1-0.8
	Saturation	D570						
	21. Dielectric strength (⅛-in. thick specimen), short time, v./mil	D149		490-520		400-450		270-400 175-215 @ 100° C.
SUPPLIERS[a]			Wilson-Fiberfil	**Allied**	Quaker Oats	**Du Pont** Schulman	**Du Pont**	**Fiberite** **Plastics Eng.** Am. Cyanamid; MKB; Patent Plastics; Perstorp; Plastics Mfg.

^{a}See Ref. 10 for suppliers of specialty materials and custom compounds.
^{b}Tensile test method varies with material: D 638 is standard for thermoplastics; D 651 for rigid thermosetting plastics; D 412 for elastomeric plastics; D 882 for thin plastics sheeting.

TABLE 4.4 (continued)

	Melamine formaldehyde (Cont'd)	Melamine phenolic	Phenolic						
			Molding compounds, phenol-formaldehyde, and furfural				Impact-modified		
	Glass fiber-reinforced	Woodflour- and cellulose-filled	Woodflour-filled	Woodflour- and mineral-filled	High-strength glass fiber-reinforced	Cotton-filled	Cellulose-filled	Fabric- and rag-filled
1.	Thermoset	Thermoset	Thermoset	Thermoset	Thermoset	Thermoset	Thermoset	Thermoset
2.	C: 280-350	C: 300-350 I: 350-400	C: 290-380 I: 330-400	C: 220-250 I: 220-250 T: 220-250	C: 300-380 I: 330-390	C: 290-380 I: 330-400	C: 290-380 I: 330-400	C: 290-380 I: 330-400 T: 300-350
3.	2-8	5-20	2-20	2-3	1-20	2-20	2-20	2-20
4.	5-10	2.1-4.4	1.0-1.5		2.0-10.0	1.0-1.5	1.0-1.5	1.0-1.5
5.	0.001-0.006	0.009-0.010	0.004-0.009	0.003-0.008	0.001-0.004	0.004-0.009	0.004-0.009	0.003-0.009
6.	5000-10,500	6000-8000	5000-9000	6500-7500	7000-18,000	6000-10,000	3500-6500	6000-8000
7.	0.6	0.4-0.8	0.4-0.8		0.2	1-2	1-2	1-4
8.								
9.	20,000-35,000	26,000-30,000	25,000-31,000	28,000	16,000-70,000	23,000-31,000	22,000-31,000	20,000-28,000
10.	14,000-23,000	8000-10,000	7000-14,000	10,000-11,000	12,200-60,000	9000-13,000	5500-11,000	10,000-14,000
11.	1600-2400	800-1700	800-1700		1900-3300	1100-1400		900-1100
12.								
13.		1000-1200	1000-1200	1200-1300	1150-33,000	800-1300	900-1300	700-1300
14.	0.6-18	0.2-0.4	0.2-0.6	0.29-0.35	0.5-18.0	0.3-1.9	0.4-1.1	0.8-3.5
15.	M115	E95-100	M100-115		E54-101	M105-120	M95-115	M105-115
16.	15-28	10-40	30-45		8-21	15-22	20-31	18-24
17.	375-400	285-310	300-370	360-380	350-600	300-400	300-350	325-400
18.	10-11.5	4-7	4-8		8-14	8-10	6-9	9-12
19.	1.5-2.0	1.5-1.7	1.37-1.46	1.44-1.56	1.69-2.0	1.38-1.42	1.38-1.42	1.37-1.45
20.	0.09-1.3	0.3-0.65	0.3-1.2	0.2-0.35	0.03-1.2	0.6-0.9	0.5-0.9	0.6-0.8
21.	130-370	220-325	260-400	330-375	140-400	200-360	300-380	200-370
	Fiberite Am. Cyanamid; U.S. Prolam	Plastics Eng.	Occidental Plastics Eng. Plaslok; Reichhold; Valite Div.	Occidental Plastics Eng. Valite Div.	Fiberite Occidental Plastics Eng. Resinoid; Rogers Corp.; Valite Div.	Fiberite Occidental Plastics Eng. Plaslok; Valite Div.	Fiberite Occidental Plaslok; Rogers Corp.; Valite Div.	Fiberite Plastics Eng. Resinoid; Rogers Corp.

(continued)

TABLE 4.4 (continued)

		Phenolic (Cont'd)				Polyamide	
		MC, phenol-formaldehyde, and furfural (Cont'd)		Casting resins		Nylon, Type 6	
		Heat resistant					
Properties	ASTM test method	Asbestos-filled	Mineral-filled	Unfilled	Mineral-filled	Molding and extrusion compound	30-35% glass fiber-reinforced
Processing							
1. Melting temperature, °C. T_m (crystalline)		Thermoset	Thermoset	Thermoset	Thermoset	210-220	210-220
T_g (amorphous)							
2. Processing temperature range, °F. (C = compression; T = transfer; I = injection; E = extrusion)		C: 290-400 I: 330-900	C: 270-350 I: 330-380			I: 440-550 E: 440-525	I: 480-550
3. Molding pressure range, 10^3 p.s.i.		2-20	2-20			1-20	3-20
4. Compression ratio		1.0-1.5	2.1-2.7			3.0-4.0	3.0-4.0
5. Mold (linear) shrinkage, in./in.	D955	0.001-0.009	0.002-0.006			0.005-0.015	0.003-0.005
Mechanical							
6. Tensile strength at break, p.s.i.	D638[b]	4500-7500	6000-9700	5000-9000	4000-9000		24,000[c]; 16,000[d]
7. Elongation at break, %	D638[b]	0.1-0.5	0.1-0.5	1.5-2.0		30-100[c]; 300[d]	3-6[c]; 6-7[d]
8. Tensile yield strength, p.s.i.	D638[b]			*		11,700[c]; 7400[d]	
9. Compressive strength (rupture or yield), p.s.i.	D695	20,000-35,000	22,500-34,600	12,000-15,000	29,000-34,000	13,000-16,000[c]	19,000-23,000[c]
10. Flexural strength (rupture or yield), p.s.i.	D790	7000-14,000	11,000-14,000	11,000-17,000	9000-12,000	15,700[c]; 5800[d]	33,000[c]; 21,000[d]
11. Tensile modulus, 10^3 p.s.i.	D638[b]	1000-3000	2400	400-700		380[c]; 100[d]	1450[c]; 800[d]
12. Compressive modulus, 10^3 p.s.i.	D695					250[d]	
13. Flexural modulus, 10^3 p.s.i. 73° F.	D790	1000-2200	1000-2000			390[c]; 140[d]	1300[c]; 800[d]
200° F.	D790						
250° F.	D790						
300° F.	D790						
14. Izod impact, ft.-lb./in. of notch (⅛-in. thick specimen)	D256A	0.26-3.5	0.26-0.36	0.24-0.4	0.35-0.5	0.6-1.0[c]; 3.0[d]	2.3-3.0[c]; 5.5[d]
15. Hardness Rockwell	D785	M105-115	E88	M93-120	M85-120	R119[c]	M96[c]; M78[d]
Shore/Barcol	D2240/D2583						
Thermal							
16. Coef. of linear thermal expansion, 10^{-6} in./in./°C.	D696	10-40	19-26	68	75	80-83	20-30
17. Deflection temperature under flexural load, °F. 264 p.s.i.	D648	300-500+	360-475	165-175	150-175	155-185[c]	410-420[c]
66 p.s.i.	D648					365-375[c]	425-430[c]
18. Thermal conductivity, 10^{-4} cal.-cm./sec.-cm.^{2}.-°C.	C177	6-22	10-14	3.5		5.8	5.8-11.4
Physical							
19. Specific gravity	D792	1.45-2.0	1.42-1.84	1.24-1.32	1.68-1.70	1.12-1.14	1.35-1.42
20. Water absorption (⅛-in. thick specimen), % 24 hr.	D570	0.1-0.5	0.1-0.3	0.1-0.36		1.3-1.9	1.1-1.2
Saturation	D570					8.5-10.0	6.5-7.0
21. Dielectric strength (⅛-in. thick specimen), short time, v./mil	D149	100-360	200-350	250-400	100-250	400[c]	400-450[c]
SUPPLIERS[a]		Fiberite Plastics Eng. Reichhold; Resinoid; Rogers Corp.	Occidental Plastics Eng. Plaslok; Rogers Corp.; Valite Div.	Union Carbide; Ametek. Haveg; Monsanto; Reichhold; Rogers Corp.	Monsanto; Reichhold; Schenectady Chem.	Allied Du Pont Schulman Adell; Custom Resins; Emser Ind.; MKB; Montedison; Nycoa; Polymer Corp.; Thermofil; Wellman	Allied RTP Adell; Emser Ind.; LNP; MKB; Nycoa; Polymer Corp.; Thermofil; Wellman; Wilson-Fiberfil

[a] See Ref. 10 for suppliers of specialty materials and custom compounds.
[b] Tensile test method varies with material: D638 is standard for thermoplastics: D651 for rigid thermosetting plastics; D412 for elastomeric plastics; D882 for thin plastics sheeting.
[c] Dry, as molded (approximately 0.2% moisture content).
[d] As conditioned to equilibrium with 50% relative humidity.

TABLE 4.4 (continued)

Polyamide (Cont'd)

	Nylon, Type 6 (Cont'd)					Nylon, Type 6/6		
	40% mineral- and glass fiber-reinforced	High-impact copolymer	High-impact; 35% mineral-filled	EMI shielding (conductive); 30% PAN carbon fiber	Cast	Molding compound	Impact-modified	30-33% glass fiber-reinforced
1.	210-220	210-220	215	210-220	216	265	240-260	265
2.	I: 480-550	I: 450-580 E: 450-550	I: 550-600	I: 540-575		I: 520-620	I: 520-580	I: 520-575
3.	2-20	1-20	5-10			1-20	1-20	5-20
4.	3.0-4.0	3.0-4.0	3			3.0-4.0		3.0-4.0
5.	0.004-0.006	0.008-0.018	0.003-0.004	0.001-0.003		0.008-0.015	0.013-0.018	0.004-0.006
6.	16,000-16,500^c	7500-11,000^c	14,000^c	30,000^c	11,000-14,000^c	12,000^c; 11,000^d	7500^c; 5800^d	28,000^c; 22,000^d
7.	2-3^c; 2-5^d	150-270^c	15^c	3^c	30-75^c	60^c; 300^d	4-80^c; 150-300^d	3-4^c; 5-7^d
8.				12,000^c		8000^c; 6500^d		
9.	18,000^c					15,000^c (yld.)		24,000-29,400^c
10.	23,000-23,200^c	5000-12,000^c	20,000^c	46,000^c	7000-17,500^c	17,000^c; 6100^d		41,000^c; 25,000^d
11.	1200^c			2800^c		350-450^d		
12.								
13.	1100-1300^c	110-320^c	725-750^c	2500^c	450^c	420^c; 185^d	245-275^c; 125-150^d	1300^c; 800^d
		60-130^c			160^c			
14.	0.6-0.9^c	1.8-No break^c	1.0^c	2.8^c	0.8-3.0^c	0.8-1.0^c; 2.1^d	3-No break^c; 2.5-No break^d	2.0-2.2^c; 2.6-3.0^d
15.	R118^c	R81-110^c			R95-120^c	R120^c; M83^c	R114-115^c	M100^c
16.	30	72			90	80		15-20
17.	390-405^c	113-130^c		415		167^c	158-160^c	485-490^c
	420-425^c	260-350^c		425	400-425^c	474^c	420-440^c	495-500^c
18.						5.8		5.1-11.7
19.	1.46-1.50	1.08-1.17	1.43	1.28	1.13-1.15	1.13-1.15	1.08-1.10	1.37-1.38
20.	0.9	1.3-1.5			1.0	0.6-1.2	1.0-1.3	0.9-1.0
	6.0	8.5				8.5-10.0	8.5	6.5
21.	500-550^c	450^c			300-400^c	600^c		440^c
	Allied Adell; Polymer Corp.; Thermofil; Wellman; Wilson-Fiberfil	Allied Adell; Custom Resins; Emser Ind.; MKB; Montedison; Nycoa; Polymer Corp.; Wellman	Allied	Thermofil; Wilson-Fiberfil	Polymer Corp.	Allied Celanese; Du Pont Schulman Adell; Belding Chem.; Emser Ind.; MKB; Monsanto; Montedison; Polymer Corp.; Thermofil; Wellman; Wilson-Fiberfil	Du Pont Adell	Celanese; Du Pont RTP Adell; LNP; MKB; Monsanto; Montedison; Polymer Corp.; Thermofil; Wellman; Wilson-Fiberfil

(continued)

66 Materials and Material Selection

TABLE 4.4 (continued)

			Polyamide (Cont'd)					
			Nylon, Type 6/6 (Cont'd)					
						EMI shielding (conductive)		
Materials	Properties	ASTM test method	40% glass- and mineral-reinforced	Mineral-filled	Modified high-impact, 25% mineral-filled	30% graphite fiber	40% aluminum flake	Anti-friction molybdenum disulfide-filled
Processing	1. Melting temperature, °C. T_m (crystalline)		255-260	265	250-260	265	265	265
	T_g (amorphous)							
	2. Processing temperature range, °F. (C = compression; T = transfer; I = injection; E = extrusion)		I: 510-590	I: 520-580	I: 510-570	I: 500-575	I: 525-600	I: 500-600
	3. Molding pressure range, 10^3 p.s.i.		9-20	5-20	10-20	10-20	10-20	5-25
	4. Compression ratio		3-4	3.0-4.0				
	5. Mold (linear) shrinkage, in./in.	D955	0.002-0.005	0.012-0.022	0.014-0.018	0.002-0.003	0.005	0.007-0.018
Mechanical	6. Tensile strength at break, p.s.i.	D638[b]	15,500-20,000[c]	14,000[c]; 11,000[d]	11,000[c]	32,000-35,000[c]	6000[c]	13,700[c]
	7. Elongation at break, %	D638[b]	2-5[c]	7[c]; 16[d]	15[c]	2-4[c]	4[c]	15[c]
	8. Tensile yield strength, p.s.i.	D638[b]						
	9. Compressive strength (rupture or yield), p.s.i.	D695	18,000-37,000[c]	15,500[c]		27,000[c]	7500[c]	12,500[c]
	10. Flexural strength (rupture or yield), p.s.i.	D790	24,000-28,900[c]	22,000[c]; 9000[d]	20,000[c]	45,000-51,000[c]	11,700[c]	17,000[c]
	11. Tensile modulus, 10^3 p.s.i.	D638[b]	1000[c]	900[c]; 500[d]		3200[c]	720[c]	550[c]
	12. Compressive modulus, 10^3 p.s.i.	D695	370[c]					
	13. Flexural modulus, 10^3 p.s.i. 73° F.	D790	985-1370[c];600[d]	900[c]; 400[d]	600[c]	2500-2900[c]	690[c]	450[c]
	200° F.	D790	600[c]					
	250° F.	D790						
	300° F.	D790						
	14. Izod impact, ft.-lb./in. of notch (⅛-in. thick specimen)	D256A	0.6-1.1[c]	1.4[c]; 3.9[d]	1.0[c]	1.5[c]	2.5[c]	4.5[c]
	15. Hardness Rockwell	D785	M95-98[c]	R106-119[c]	R120	R120[c]	R114[c]	R119[c]
	Shore/Barcol	D2240/ D2583						
Thermal	16. Coef. of linear thermal expansion, 10^{-6} in./in./°C.	D696	20-29	27	30	11	22	54
	17. Deflection temperature under flexural load, °F. 264 p.s.i.	D648	432-475[c]	300-360[c]	320	495	380	260[c]
	66 p.s.i.	D648	480-496[c]	320-460[c]	399	500	400	
	18. Thermal conductivity, 10^{-4} cal.-cm./ sec.-cm.^{2}.°C.	C177	11	9.6		24.1		
Physical	19. Specific gravity	D792	1.42-1.49	1.39-1.47	1.28	1.28-1.43	1.48	1.15-1.17
	20. Water absorption (⅛-in. thick specimen), % 24 hr.	D570	0.4-0.9	0.6-0.9	1.1	0.5	1.1	0.8-1.1
	Saturation	D570		6.0-6.5				
	21. Dielectric strength (⅛-in. thick specimen), short time, v./mil	D149	300-525	450[c]				360[c]
	SUPPLIERS[a]		Celanese; Du Pont; Adell; Monsanto; Polymer Corp.; Thermofil; Wellman; Wilson-Fiberfil	Celanese; Du Pont; Adell; Monsanto; Polymer Corp.; Thermofil; Wellman; Wilson-Fiberfil	Wellman	LNP; Thermofil; Wilson-Fiberfil	Thermofil; Wilson-Fiberfil	Adell; LNP; Polymer Corp.; Thermofil; Wilson-Fiberfil

[a]See Ref. 10 for suppliers of specialty materials and custom compounds.
[b]Tensile test method varies with material: D638 is standard for thermoplastics; D651 for rigid thermosetting plastics; D412 for elastomeric plastics; D882 for thin plastics sheeting.
[c]Dry, as molded (approximately 0.2% moisture content).
[d]As conditioned to equilibrium with 50% relative humidity.

TABLE 4.4 (continued)

#			Nylon, Type 6/12		Nylon, Type 11	Nylon, Type 12	Aromatic polyamide	
	Nylon, Type 6/6-6 copolymer	Nylon, Type 6/9 molding and extrusion	Molding compound	30-35% glass fiber-reinforced	Molding and extrusion compound	Molding and extrusion compound	Amorphous transparent copolymer	Aramid molded parts, unfilled
1.	240	205	195-217	215-217	191-194	176-179		275
							125-155	
2.	I: 450-500	I: 450-550 E: 425-500	I: 450-550 E: 464-469	I: 450-550	I: 390-520 E: 390-475	I: 360-525 E: 350-500	I: 480-610 E: 520-595	
3.	1-15	1-15	1-15	4-20	1-15	1-15	5-20	
4.						2.5-4		
5.	0.006-0.015	0.010-0.015	0.011	0.003-0.005	0.012	0.003-0.015	0.004-0.007	
6.	7400-12,400^c	8500^c	8800^c	22,000^c; 20,000^d	8000^c	8000-9000^c; 8000^d	7600-14,000^c; 13,000^d	17,500^c
7.	40^c;300^d	1125^c	150^c; 300^d	4^c; 5^d	300^c	250-350^c	40-150^c; 260^d	5^c
8.			5800^c; 3100^d			5800-6100^c	11,000^c; 11,000^d	
9.				22,000^c			17,500^c; 14,000^d	30,000^c
10.			11,000^c; 4300^d	32,000^c		7600-8100^c	10,000-16,400^c; 14,000^d	25,800^c
11.	150-410^c	275^c	218-290^c; 123-180^d	1200^c; 900^d	185^c	180^c	275-410^c; 270^d	
12.						180^c	340^c	290^c
13.	150-410^c	290^c	240-290^c; 74-100^d	1100^c; 900^d	150^c	120-180^c	306-400^c; 350^d	640^c
14.	0.7^c; No break^d	1.1^c	1.0-1.9^c; 1.4-No break^d	2.4^c; 2.8^d	1.8^c	1.6-5.5^c	1.0-3.5^c; 1.8-2.7^d	1.4^c
15.	R119^c; R83^d	R111^c	M78^c; M34^d	M93^c; E40-50^d	R108^c	R106-109^c; 105^d	M77-93^c	E90^c
			D80^c; D63^d			D72-73	D83^c; D85^d	
16.				100	100	28-70	40	
17.	170^c	135-140^c	180^c	415-425^c	130^c	126-131^c	248-256^c	500^c
	430^c	330-340^c	330^c	400-430^c	300^c	293^c	261-285^c	
18.			5.2	10.2	8	5.2-7.3	5	5.2
19.	1.08-1.14	1.08-1.10	1.06-1.10	1.30-1.38	1.03-1.05	1.01-1.02	1.06-1.19	1.30
20.	1.5-2.0	0.5	0.4	0.2	0.3	0.25-0.30	0.4	0.6
	9.0-10.0		2.5-3.0	1.85		0.75-0.9	1.3-4.2	
21.	400^c	600^c	400^c	520^c	425^c	450^c	350^c	800^c
	Du Pont; Emser Ind.; Monsanto	Belding Chem.; Monsanto	Du Pont; Emser Ind.; Polymer Corp.	Du Pont; LNP; Wilson-Fiberfil	Rilsan	Huels; Rilsan; Emser Ind.	Allied; Du Pont; Union Carbide; Upjohn; Emser Ind.; Kay-Fries	Du Pont

(continued)

TABLE 4.4 (continued)

			Polyamide-imide			Polyaryl ether	Polybutadiene	Polybutylene
	Properties	ASTM test method	Unfilled compression and injection molding compound	30% glass fiber-reinforced	Graphite fiber-reinforced	Unfilled	Casting resin	Extrusion compound
Processing	1. Melting temperature, °C: T_m (crystalline)						Thermoset	126
	T_g (amorphous)		275	275	275	160		
	2. Processing temperature range, °F (C = compression, T = transfer; I = injection; E = extrusion)		C: 600-650 I: 600-650	C: 630-650 I: 600-675	C: 630-650 I: 600-675	I: 540-590		C: 300-350 I: 290-380 E: 290-380
	3. Molding pressure range, 10^3 p.s.i.		2-40	15-40	15-40	10-20		10-30
	4. Compression ratio		1.0-1.5	1.0-1.5	1.0-1.5	1.8-2.6		2.5
	5. Mold (linear) shrinkage, in./in.	D955	0.006-0.008	0.003-0.005	0.000-0.002	0.004-0.007		0.003 (unaged) 0.026 (aged)
Mechanical	6. Tensile strength at break, p.s.i.	D638[b]	17,000-26,900	28,300	29,800	7500		3800-4400
	7. Elongation at break, %	D638[b]	12-15	5-6	6-7	80		300-380
	8. Tensile yield strength, p.s.i.	D638[b]						1700-2500
	9. Compressive strength (rupture or yield), p.s.i.	D695	35,000-40,000					
	10. Flexural strength (rupture or yield), p.s.i.	D790	27,400-30,700	46,000	45,900	11,000	8000-14,000	2000-2300
	11. Tensile modulus, 10^3 p.s.i.	D638[b]	600-730	1670	2900	320	560	30-40
	12. Compressive modulus, 10^3 p.s.i.	D695	280-450	520	530			31
	13. Flexural modulus, 10^3 p.s.i. 73°F	D790	520-665	1610	2600	300		45-50
	200°F	D790	440-590					
	250°F	D790	420-560					
	300°F	D790	400-520	1220				
	14. Izod impact, ft.-lb./in. of notch (1/8-in. thick specimen)	D256A	2.5-4.0	2.0	1.2	8.0		No break
	15. Hardness Rockwell	D785	E78	E94	E94	R117	R40	
	Shore/Barcol	D2240/ D2583						Shore D55-65
	16. Coef. of linear thermal expansion, 10^{-6} in./in./°C	D696	36	18	11	65		128-150
Thermal	17. Deflection temperature under flexural load, °F 264 p.s.i.	D648	500-525	525	525	300		130-140
	66 p.s.i.	D648				320		215-235
	18. Thermal conductivity, 10^{-4} cal.-cm./ sec.-cm.^{2}·°C	C177	5.6-5.8			7.14		5.2
Physical	19. Specific gravity	D792	1.38-1.40	1.57	1.41	1.14	0.97	0.91-0.925
	20. Water absorption (1/8-in. thick specimen), % 24 hr.	D570				0.25	0.03	0.01-0.02
	Saturation	D570	0.28	0.22				
	21. Dielectric strength (1/8-in. thick specimen), short time, v./mil	D149	440-600	560		430	630	~450
	SUPPLIERS[a]		Amoco	Amoco LNP	Amoco	Uniroyal	Occidental Colorado Chem.	Shell

[a]See Ref. 10 for suppliers of specialty materials and custom compounds.
[b]Tensile test method varies with material: D638 is standard for thermoplastics; D651 for rigid thermosetting plastics; D412 for elastomeric plastics; D882 for thin plastics sheeting.

TABLE 4.4 (continued)

Polycarbonate							
Unfilled molding and extrusion resins		Glass fiber-reinforced				EMI shielding (conductive)	
High-viscosity	Low-viscosity	10% glass	30% glass	Polyester copolymer	Modified polycarbonate blends	30% graphite fiber	40% PAN carbon fiber
1.							
150	140	150	150	170-180		149	
2. I: 560	I: 520	I: 520-650	I: 560-650	I: 600-710	I: 475-560	I: 540-650	I: 580-620
3. 10-20	8-15	10-20	10-30	8-20			
4. 1.74-5.5	1.74-5.5			1.5-3	2-2.5		
5. 0.005-0.007	0.005-0.007	0.002-0.005	0.001-0.002	0.007-0.010	0.005-0.009	0.001-0.002	0.0005
6. 9500	9500	9500	19,000	10,000-11,300	8000-8500	24,000	24,000
7. 110	110	5	3-5	65-122	120-145	2-5	1.6
8. 9000	9000			9500-9800	7400-8300		
9. 12,500	12,500	13,500	18,000	11,500		26,000	
10. 13,500	13,500	15,000	23,000	13,000-13,800	12,000-12,500	34,000-36,000	35,000
11. 345	345	500	1250			2150	3000
12. 350	350	520	1300				
13. 340	340	500	1100	294-338	280-325	1900	2800
275	275	440	960				
245	245	420	900				
14. 16 @ ¼ in.	14 @ ¼ in.	1.2	2.0	6.0-10	2-15	1.8	2.0
15. M70	M70	M75	M92	M82-92	R114-122	R118	
16. 68	68	38	22	81-92	80-95	9	
17. 270	270	288	295	305-325	203-250	280-300	295
280	280		305	320-345	223-265	295	300
18. 4.7	4.7	4.6	5.2	4.9-5.0	4.3	16.9	
19. 1.2	1.2	1.27-1.28	1.4	1.2	1.20-1.22	1.33	1.38
20. 0.15	0.15	0.15	0.14	0.16-0.19	0.12-0.14	0.04-0.08	0.13
					0.35-0.60		
21. 380	380	530	475	509-520	440-500		
Shuman Dow Chem.; GE. Pittsfield; Mobay	**Shuman** Dow Chem.; GE. Pittsfield; Mobay	GE. Pittsfield; LNP; Mobay; Thermofil; Wilson-Fiberfil	GE. Pittsfield; LNP; Mobay; Thermofil; Wilson-Fiberfil	GE. Pittsfield; Mobay	GE. Pittsfield	LNP; Thermofil	Wilson-Fiberfil

(continued)

Materials and Material Selection

TABLE 4.4 (continued)

	Properties	ASTM test method	PBT Unfilled	PBT 30% glass fiber-reinforced	PBT 40-45% glass fiber- and mineral-reinforced	PBT 35% glass fiber- and mica-reinforced	PBT EMI shielding (conductive); 30% carbon fiber	PET Unfilled
Processing	1. Melting temperature, °C. T_m (crystalline)		232-267	232-267	220-228	220-224	222	254-259
	T_g (amorphous)							73
	2. Processing temperature range, °F. (C = compression; T = transfer; I = injection; E = extrusion)		I: 435-525	I: 440-530	I: 450-500	I: 480-510	I: 430-550	I: 540-600
	3. Molding pressure range, 10^3 p.s.i.		4-10	5-15	10-15	9-15	5-20	2-7
	4. Compression ratio				3-4			3.1
	5. Mold (linear) shrinkage, in./in.	D955	0.015-0.020	0.002-0.008	0.003-0.010	0.003-0.012	0.001-0.004	0.020-0.025
Mechanical	6. Tensile strength at break, p.s.i.	D638[b]	8200	17,000-19,000	12,000-14,600	11,400-13,800	22,000-23,000	7000-10,500
	7. Elongation at break, %	D638[b]	50-300	2-4	2-5	2-3	1-3	50-300
	8. Tensile yield strength, p.s.i.	D638[b]						
	9. Compressive strength (rupture or yield), p.s.i.	D695	8600-14,500	18,000-23,500	15,000			11,000-15,000
	10. Flexural strength (rupture or yield), p.s.i.	D790	12,000-16,700	26,000-29,000	18,500-23,000	18,000-22,000	29,000-34,000	14,000-18,000
	11. Tensile modulus, 10^3 p.s.i.	D638[b]	280	1300	1350		3500	400-600
	12. Compressive modulus, 10^3 p.s.i.	D695						
	13. Flexural modulus, 10^3 p.s.i. 73° F.	D790	330-400	1100-1200	1250-1600	1200-1400	2300-2700	350-450
	200° F.	D790						
	250° F.	D790						
	300° F.	D790						
	14. Izod impact, ft.-lb./in. of notch (⅛-in. thick specimen)	D256A	0.8-1.0	1.3-1.6	0.7-1.7	1.3-1.8	1.2-1.4	0.25-0.65
	15. Hardness Rockwell	D785	M68-78	M90	M75-86	M50		M94-101
	Shore/Barcol	D2240/ D2583						
Thermal	16. Coef. of linear thermal expansion, 10^{-6} in./in./°C.	D696	60-95	25				65
	17. Deflection temperature under flexural load, °F. 264 p.s.i.	D648	122-185	428	390-395	330-387	420-430	100-106 (annealed)
	66 p.s.i.	D648	240-375	437	426	410-416		
	18. Thermal conductivity, 10^{-4} cal.-cm./ sec.-cm.^{2}.-°C.	C177	4.2-6.9	7.0			15.8	3.3-3.6
Physical	19. Specific gravity	D792	1.31-1.38	1.52	1.65-1.74	1.59-1.73	1.41-1.42	1.34-1.39
	20. Water absorption (⅛-in. thick specimen), % 24 hr.	D570	0.08-0.09	0.06-0.08	0.04-0.05	0.06-0.07	0.04-0.05	0.1-0.2
	Saturation	D570						
	21. Dielectric strength (⅛-in. thick specimen); short time, v./mil	D149	420-550	460-550	540-590	450-600		
SUPPLIERS[a]			Celanese. Dainippon; GAF; GE, Pittsfield; Montedison; Thermofil	Celanese. GAF; GE, Pittsfield; LNP; Montedison; Thermofil; Wilson-Fiberfil	Celanese. GE, Pittsfield; Thermofil	GAF; GE, Pittsfield	LNP; Wilson-Fiberfil	Amoco; Du Pont; Eastman. Schulman. Goodyear

[a]See Ref. 10 for suppliers of specialty materials and custom compounds.
[b]Tensile test method varies with material: D 638 is standard for thermoplastics; D 651 for rigid thermosetting plastics; D 412 for elastomeric plastics; D 882 for thin plastics sheeting.

TABLE 4.4 (continued)

	Polyester, thermoplastic (Cont'd)					Polyester, thermosetting and alkyd			
	PET (Cont'd)			Polyarylate		Cast		Glass fiber-reinforced	
	30% glass fiber-reinforced	40-45% glass fiber- and mica-reinforced	EMI shielding (conductive); 30% PAN carbon fiber	Transparent copolymer	Injection molding and extrusion	30% glass fiber-reinforced	Rigid	Flexible	Preformed, chopped roving	Premix, chopped glass
1.	245-254	252-255		265			Thermoset	Thermoset	Thermoset	Thermoset
				81	190	190				
2.	I: 520-590	I: 540-580	I: 550-590	I: 450-525 / E: 490-550	I: 600-750 / E: 600-650	I: 600-640			C: 170-320	C: 280-350
3.	4-20	5-20			5-20	16-19			0.25-2	0.5-2
4.	2-3	4		3.6	2.5-3	2.5-3				
5.	0.002-0.009	0.002-0.004	0.001-0.002		0.006-0.009	0.001			0.0002-0.002	0.001-0.012
6.	21,000-23,000	14,000-17,300	25,000	4000	8800-9700	21,600	6000-13,000	500-3000	15,000-30,000	3000-10,000
7.	2-7	2-3	1.4	180	50-65		<2	40-310	1-5	<1
8.	23,000			7100	9900-10,000					
9.	25,000	20,500-24,000			12,000	24,600	13,000-30,000		15,000-30,000	20,000-30,000
10.	32,000-33,500	21,000-26,700	38,000	10,000	11,000-14,500	32,000	8500-23,000		10,000-40,000	7000-20,000
11.	1300-1440	1950	3600		300-330		300-640		800-2000	1000-2500
12.										
13.	1250-1500	1400-1750	2700	290	310-330	1120			1000-3000	1000-2000
	520	489			260-290					
					255-285					
	390	320			275					
14.	1.6-1.9	1.2-1.4	1.5	1.7	4.1-5.5	2.7	0.2-0.4	>7	2-20	1.5-16
15.	M100	R118		R105	M85-100					
							Barcol 50-75	Shore D84-94	Barcol 50-80	Barcol 50-80
16.	25-30	21		51	62-63		55-100		20-50	20-33
17.	410-435	412-420	430	145	338-345	355	140-400		>400	>400
	430	420	470	158	355					
18.	6.0-6.9			7.8	4.3					
19.	1.56-1.67	1.58-1.65	1.42	1.27	1.21-1.22	1.44	1.10-1.46	1.01-1.20	1.35-2.30	1.65-2.30
20.	0.05	0.05	0.05	0.07	0.26-0.27		0.15-0.6	0.5-2.5	0.01-1.0	0.06-0.28
							0.5-0.71			
21.	430-650	570-600		400	400-465	515	380-500	250-400	350-500	345-420
	Allied Du Pont; Schulman LNP; Mobay; Thermofil; Wilson-Fiberfil	Du Pont; Mobay; Thermofil; Wilson-Fiberfil	Wilson-Fiberfil	Eastman	Occidental Union Carbide	Occidental	ICI Americas USS Chem. Cargill; Freeman; Owens-Corning; Reichhold; Shell	ICI Americas USS Chem. Cargill; Freeman; Owens-Corning; Reichhold; Shell	Eagle-Picher; Glastic; Haysite; Koppers; Plumb; Premix; Rostone	Am. Cyanamid; Eagle-Picher; Glastic; Haysite; Koppers; Plumb; Premix; Rostone

(continued)

TABLE 4.4 (continued)

| | | | Polyester, thermosetting and alkyd (Cont'd) | | | | |
| | | | Glass fiber-reinforced (Cont'd) | | | | EMI shielding (conductive) | |
Properties	ASTM test method	Woven cloth	SMC	SMC, low-shrink	BMC, TMC	SMC, TMC	BMC
Processing							
1. Melting temperature, °C T_m (crystalline)		Thermoset	Thermoset	Thermoset	Thermoset	Thermoset	Thermoset
$\quad$ T_g (amorphous)							
2. Processing temperature range, °F. (C = compression; T = transfer; I = injection; E = extrusion)		C: 73-250	C: 270-350	C: 270-330 I: 270-330	C: 310-380 I: 300-370 T: 280-320	C: 270-380 I: 270-370 T: 280-320	C: 310-380 I: 300-370 T: 280-320
3. Molding pressure range, 10^3 p.s.i.		0.3	0.3-1.2	0.5-2		0.5-2	
4. Compression ratio				1.0		1.0	
5. Mold (linear) shrinkage, in./in.	D955	0.0002-0.002	0.001-0.004	0.0002-0.001	0.0005-0.004	0.0002-0.001	0.0005-0.004
Mechanical							
6. Tensile strength at break, p.s.i.	D638[b]	30,000-50,000	8000-25,000	4500-20,000	5000-10,000	7000-8000	4000-4500
7. Elongation at break, %	D638[b]	1-2	3	3-5			
8. Tensile yield strength, p.s.i.	D638[b]						
9. Compressive strength (rupture or yield), p.s.i.	D695	25,000-50,000	15,000-30,000	15,000-30,000	14,000-30,000	20,000-24,000	18,000
10. Flexural strength (rupture or yield), p.s.i.	D790	40,000-80,000	10,000-36,000	9000-35,000	16,000-24,000	18,000-20,000	12,000
11. Tensile modulus, 10^3 p.s.i.	D638[b]	1500-4500	1400-2500	1000-2500	1500-2500		
12. Compressive modulus, 10^3 p.s.i.	D695						
13. Flexural modulus, 10^3 p.s.i. $\quad$ 73° F.	D790	1000-3000	1000-2200	1000-2500		1400-1500	1400-1500
$\qquad$ 200° F.	D790	660					
$\qquad$ 250° F.	D790	430					
$\qquad$ 300° F.	D790	270					
14. Izod impact, ft.-lb./in. of notch (⅛-in. thick specimen)	D256A	5-30	7-22	2.5-15	4-13	10-12	5-7
15. Hardness $\quad$ Rockwell	D785						
$\qquad$ Shore/Barcol	D2240/ D2583	Barcol 60-80	Barcol 50-70	Barcol 40-70	Barcol 50-60	Barcol 45-50	Barcol 50
Thermal							
16. Coef. of linear thermal expansion, 10^{-6} in./in./°C.	D696	15-30	20	6-30			
17. Deflection temperature under flexural load, °F. $\quad$ 264 p.s.i.	D648	>400	375-500	375-500	320-350	395-400 +	400 +
$\qquad$ 66 p.s.i.	D648						
18. Thermal conductivity, 10^{-4} cal.-cm./ sec.-cm.^{2}-°C.	C177				18-22		
Physical							
19. Specific gravity	D792	1.50-2.10	1.65-2.6	1.6-2.4	1.72-2.05	1.75-1.80	1.80-1.85
20. Water absorption (⅛-in. thick specimen), % $\quad$ 24 hr.	D570	0.05-0.5	0.1-0.25	0.01-0.25	0.1-0.45		
$\qquad$ Saturation	D570					0.5	0.5
21. Dielectric strength (⅛-in. thick specimen), short time, v./mil	D149	350-500	380-500	380-450	300-390		
SUPPLIERS[a]		Eagle-Picher; Glastic; Haysite; Koppers; MKB; Plumb; Premix; Rostone	Armco; Budd; Haysite; Premix; Rostone	Armco; Budd; Haysite; Premix; Rostone	Plastics Eng.; Epic Resins; Glastic; Haysite; Ind. Dielectrics; Plumb; Premix; Rostone	Premix	Premix

[a] See Ref. 10 for suppliers of specialty materials and custom compounds.
[b] Tensile test method varies with material: D638 is standard for thermoplastics; D651 for rigid thermosetting plastics; D412 for elastomeric plastics; D882 for thin plastics sheeting.

TABLE 4.4 (continued)

Polyester, thermosetting and alkyd (Cont'd)		Polyether-etherketone		Polyetherimide			Polyethylene and ethylene copolymers	
Alkyd molding compounds							Low and medium density	
							Polyethylene homopolymers	
Granular and putty, mineral-filled	Glass fiber-reinforced	Unfilled	30% glass fiber-reinforced	Unfilled	30% glass fiber-reinforced	EMI shielding (conductive); 30% carbon fiber	Branched	Linear
1. Thermoset	Thermoset	334	334				106-115	122-124
				215	215	215		
2. C: 270-350 / I: 280-390 / T: 320-360	C: 290-350 / I: 280-380	I: 680-750 / E: 660-720	I: 680-750	I: 640-800	I: 620-800	I: 640-780	I: 300-450 / E: 250-450	I: 350-500 / E: 450-600
3. 2-20	2-25	10-20	10-20	10-18	10-20	10-20	5-15	5-15
4. 1.8-2.5	1-11	3	3	1.5-3	1.5-3	1.5-3	1.8-3.6	
5. 0.003-0.010	0.001-0.010	0.011	0.002	0.005-0.007	0.002	0.0005-0.002	0.015-0.050	
6. 3000-9000	4000-9500	10,200	23,500		25,000-28,500	29,000-34,000	1200-4550	1900-4000
7.		50-150	3	60	3-5	1.4-3	100-650	100-950
8.		13,200		15,200	24,500		1300-2100	1400-2800
9. 12,000-38,000	15,000-36,000		22,400	20,300	23,500-24,000	32,000		
10. 6000-17,000	8500-26,000		34,400	21,000	33,000-37,000	37,000-44,000		
11. 500-3000	2000-2800		1250	430	1300	2600-2900	25-41	38-75
12. 2000-3000				420	550			
13. 2000	2000	560	1260	480	1200-1250	2500	35-48	40-75
		435		370	1100			
				360	1060			
		290		350	1040			
14. 0.3-0.5	0.5-16	1.6	2.7	1.0	1.7-2.0	1.2-1.6	No break	1.0-9.0
15. E98	E95			M109	M125	M127		
					11	9	Shore D44-50	
16. 20-50	15-33			56	20		100-220	
17. 350-500	400-500	320	594	392	408-420	405-420		
				410	412-414	410-415	104-112	
18. 12-25	15-25			1.6	9.3	17.6	8	
19. 1.6-2.3	2.0-2.3	1.30-1.32	1.51	1.27	1.49-1.51	1.39-1.42	0.917-0.932	0.918-0.935
20. 0.05-0.5	0.03-0.5	0.1	0.06	0.25	0.18-0.20	0.2	<0.01	
		0.5			1.25	0.9		
21. 350-450	250-530			480	495-630		450-1000	
Occidental Plastics Eng. Am. Cyanamid; Plumb	Occidental Plastics Eng. Am. Cyanamid; Plaskon; Plumb; U.S. Prolam	ICI Americas	ICI Americas	GE, Pittsfield	GE, Pittsfield; LNP; Thermofil	LNP; Thermofil; Wilson-Fiberfil	Bamberger Eastman El Paso Monmouth; Union Carbide; USI Chemplex; Dow Chem.; Exxon; Gulf; Northern Petrochemical; Soltex; Washington Penn	Bamberger Du Pont, Canada Union Carbide; Dow Chem.; Soltex

(continued)

TABLE 4.4 (continued)

	Properties	ASTM test method	Polyethylene and ethylene copolymers (Cont'd)					
			Low and medium density (Cont'd)		High density			
			Ethylene copolymers				Copolymers	
			Ethylene vinyl-acetate	Ethylene ethyl-acrylate	Polyethylene homo-polymer	Rubber-modified	Low and medium molecular weight	High molecular weight
Processing	1. Melting temperature, °C T_m (crystalline)		103-108		130-137	122-127	125-132	125-133
	T_g (amorphous)							
	2. Processing temperature range, °F. (C = compression; T = transfer; I = injection; E = extrusion)		C: 200-300 I: 350-430 E: 300-380	C: 200-300 I: 250-500	I: 350-500 E: 350-525	E: 360-450	I: 375-500 E: 300-500	I: 375-500 E: 375-475
	3. Molding pressure range, 10^3 p.s.i.		1-20	1-20	12-15		5-20	
	4. Compression ratio				2		2	
	5. Mold (linear) shrinkage, in./in.	D955	0.007-0.035	0.015-0.035	0.015-0.040		0.012-0.040	0.015-0.040
Mechanical	6. Tensile strength at break, p.s.i.	D638[b]	2200-4000	1600-2100	3200-4500	2300-2900	3000-6500	2500-4300
	7. Elongation at break, %	D638[b]	300-750	700-750	10-1200	600-700	10-1300	170-800
	8. Tensile yield strength, p.s.i.	D638[b]	1200-6000		3800-4800	1400-2600	2600-4200	2800-3900
	9. Compressive strength (rupture or yield), p.s.i.	D695		3000-3600	2700-3600		2700-3600	
	10. Flexural strength (rupture or yield), p.s.i.	D790						
	11. Tensile modulus, 10^3 p.s.i.	D638[b]	7-29	4-7.5	155-158		90-130	136
	12. Compressive modulus, 10^3 p.s.i.	D695						
	13. Flexural modulus, 10^3 p.s.i. 73° F.	D790	7.7		145-225		120-180	125-175
	200° F.	D790						
	250° F.	D790						
	300° F.	D790						
	14. Izod impact, ft.-lb./in. of notch (1/8-in. thick specimen)	D256A	No break	No break	0.4-4.0		0.35-6.0	3.2-4.5
	15. Hardness Rockwell	D785						
	Shore/Barcol	D2240/ D2583	Shore D17-45	Shore D27-38	Shore D66-73	Shore D55-60	Shore D58-70	Shore D63-65
Thermal	16. Coef. of linear thermal expansion, 10^{-6} in./in./°C.	D696	160-200	160-250	59-110		70-110	70-110
	17. Deflection temperature under flexural load, °F. 264 p.s.i.	D648						
	66 p.s.i.	D648			175-196		149-176	154-158
	18. Thermal conductivity, 10^{-4} cal.-cm./ sec.-cm.^{2}-°C.	C177			11-12		10	
Physical	19. Specific gravity	D792	0.922-0.943	0.93	0.952-0.965	0.932-0.939	0.939-0.960	0.947-0.955
	20. Water absorption (1/8-in. thick specimen), % 24 hr.	D570	0.05-0.13	0.04	<0.01		<0.01	
	Saturation	D570						
	21. Dielectric strength (1/8-in. thick specimen), short time, v./mil	D149	620-760	450-550	450-500		450-500	
	SUPPLIERS[a]		Du Pont El Paso Union Carbide; USI Chemplex; Exxon; Gulf; Northern Petrochemical	Union Carbide	Allied Amoco; Bamberger Du Pont Monmouth; Schulman Shuman Union Carbide; USI Am. Hoechst; Chemplex; Dow Chem.; Gulf; Phillips; Soltex	Allied Soltex	Allied Amoco; Bamberger Du Pont Schulman Union Carbide; Am. Hoechst; Chemplex; Dow Chem.; Gulf; Phillips; Soltex	Allied Am. Hoechst; Chemplex; Phillips; Soltex

[a]See Ref. 10 for suppliers of specialty materials and custom compounds.
[b]Tensile test method varies with material: D638 is standard for thermoplastics; D651 for rigid thermosetting plastics; D412 for elastomeric plastics; D882 for thin plastics sheeting.

TABLE 4.4 (continued)

	Polyethylene and ethylene copolymers (Cont'd)					Polyimide			Poly-methyl-pentene
	High density (Cont'd)			Crosslinked		Thermoplastic			
	Copolymers (Cont'd)								
	Ethylene methyl-acrylate	Ultra high molecular weight	30% glass fiber-reinforced	Molding grade	Wire and cable grade	Unfilled	40% graphite-filled	Thermoset, 50% glass fiber-reinforced	Unfilled
1.	83	125-135	120-140					Thermoset	230-235
						310-365	365		
2.	E: 300-350	C: 400-500	I: 400-600	C: 240-450 I: 250-300	E: 250-300	C: 625-690	C: 690	C: 460 I: 390 T: 390	C: 510-550 I: 510-610 E: 510-650
3.		5-20	10-20			3-5	3-5	3-10	1-6
4.						3-4			2.0-3.5
5.		0.040	0.002-0.006	0.007-0.090	0.020-0.050			0.002	0.012-0.030
6.	1650	5600	9000	1600-4600	1500-3100	10,500-17,100	7600	6400	2100-2500
7.	740	420-525	1.5	10-440	180-600	8-10	3		64-380
8.	1650	3100-4000				12,500			2000-3300
9.			7000	2000-5500		30,000-40,000	18,000	34,000	
10.			11,000	2000-6500		19,000-28,800	14,000	21,300	
11.	12			50-500		300			160-280
12.				50-150					114-171
13.		130-140	800	70-350	8-14	450-500	700	1980	110-260
									36
									26
									17
14.		No break	1.1	1-20		1.5	0.7	5.6	2-3
15.		R50	R75			E52-99	E27	M118	R35-85
				Shore D55-80	Shore D33-57				
16.		130	48	100	100	45-56	38	13	65
17.		110-120	250	105-145	100-173	530-680	680	660	120
		155-180	265	130-225					175-195
18.			11			2.3-2.6	41.4	8.5	4.0
19.	0.942-0.945	0.94	1.28	0.95-1.45	0.91-1.40	1.36-1.43	1.65	1.6-1.7	0.833-0.835
20.	0.0	<0.01	0.02	0.01-0.06	0.01-0.06	0.24	0.14	0.7	0.01
						1.2	0.6		0.01
21.	710	500		230-550	620-760	560			450
	Gulf	Am. Hoechst; Himont USA	Ferro; Schulman; LNP; Thermofil; Wilson-Fiberfil	USI; Phillips	Union Carbide; USI	Du Pont; Upjohn; Fluorocarbon; Monsanto; Rhone-Poulenc	Du Pont; Upjohn; Fluorocarbon; Rhone-Poulenc	Ciba-Geigy; Fiberite; Upjohn; Rhone-Poulenc	Mitsui U.S.A.

(continued)

TABLE 4.4 (continued)

Materials / Properties	ASTM test method	Polyphenylene oxide and polyphenylene ether, modified					EMI shielding (conductive)
		Low glass transition	High glass transition	Impact-modified	30% glass fiber-reinforced	Mineral-filled	30% graphite fiber
Processing							
1. Melting temperature, °C, T_m (crystalline)							
T_g (amorphous)		104-110	117-142	135	100-110	110-135	
2. Processing temperature range, °F. (C = compression; T = transfer; I = injection; E = extrusion)		I: 400-600 E: 420-500	I: 425-630 E: 460-525	I: 425-550	I: 500-630 E: 460-525	I: 540-590 E: 470-530	I: 500-600
3. Molding pressure range, 10^3 p.s.i.		12-20	12-20	10-15	15-40	12-20	10-20
4. Compression ratio		1.3-3	1.3-3			2-3	
5. Mold (linear) shrinkage, in./in.	D955	0.005-0.007	0.006-0.008	0.006	0.001-0.004	0.005-0.007	0.001
Mechanical							
6. Tensile strength at break, p.s.i.	D638[b]	7800	9600	7000-8000	17,000-18,500		18,700
7. Elongation at break, %	D638[b]	50	60	35	3-5	25	2.5
8. Tensile yield strength, p.s.i.	D638[b]	6500-7800	7000-9000			9500-11,000	
9. Compressive strength (rupture or yield), p.s.i.	D695	12,000-16,400	16,400	10,000	17,900		20,000
10. Flexural strength (rupture or yield), p.s.i.	D790	11,000-12,800	9500-14,000	8200-11,000	20,000-23,000		24,000
11. Tensile modulus, 10^3 p.s.i.	D638[b]	380	355-380	345-360	1200		1150
12. Compressive modulus, 10^3 p.s.i.	D695						
13. Flexural modulus, 10^3 p.s.i. 73° F.	D790	325-400	330-400	325-345	1100-1150	425-500	1100
200° F.	D790	260	305		1000		
250° F.	D790						
300° F.	D790						
14. Izod impact, ft.-lb./in. of notch (1/8-in. thick specimen)	D256A	4-5	5	6-7	1.7-2.3	3-4	1.3
15. Hardness Rockwell	D785	R115-116	R118-120	L108, M93	R115-116	R121	R111
Shore/Barcol	D2240/ D2583						
Thermal							
16. Coef. of linear thermal expansion, 10^{-6} in./in./°C.	D696	38-68	33-60		14-25		11
17. Deflection temperature under flexural load, °F. 264 p.s.i.	D648	180-215	225-265	190-275	275-317	190-230	240
66 p.s.i.	D648	230	279	205-245	285		265
18. Thermal conductivity, 10^{-4} cal.-cm./sec.-cm.^{2}-°C.	C177	3.8	5.2		3.8-4.1		
Physical							
19. Specific gravity	D792	1.07-1.10	1.06-1.09	1.27-1.36	1.08-1.09	1.24-1.25	1.25
20. Water absorption (1/8-in. thick specimen), % 24 hr.	D570	0.06-0.1	0.06-0.12	0.1-0.07	0.06	0.07	0.04
Saturation	D570						
21. Dielectric strength (1/8-in. thick specimen), short time, v./mil	D149	400-665	500-700	530	630	490	
SUPPLIERS[a]		Borg-Warner, GE, Selkirk	Borg-Warner, GE, Selkirk	GE, Selkirk	GE, Selkirk, LNP, Thermofil	GE, Selkirk	Thermofil

[a]See Ref. 10 for suppliers of specialty materials and custom compounds. [b]Tensile test method varies with material: D638 is standard for thermoplastics; D651 for rigid thermosetting plastics; D412 for elastomeric plastics; D882 for thin plastics sheeting.

TABLE 4.4 (continued)

#	PPO/PPE, modified (Cont'd) — EMI shielding (conductive) (Cont'd) — 40% aluminum flake	Polyphenylene sulfide — Unfilled	40% glass fiber-reinforced	Mineral- and glass-filled	EMI shielding (conductive); 30% carbon fiber	Polypropylene — Homopolymer	Copolymer	40% talc-filled homopolymer	40% calcium carbonate-filled homopolymer
1.	290	275-290	285-290	275	168	160-168	158-168	168	
		88	88	88					
2.	I: 500-600	I: 600-625	I: 600-675	I: 600-650	I: 500-675	I: 400-550 E: 400-500	I: 400-550 E: 400-500	I: 410-550	I: 375-450
3.	10-20	5-15	5-20	5-20	5-20	10-20	10-20	10-20	8-14
4.						2.0-2.4	2.0-2.4		
5.	0.001	0.006-0.008	0.002-0.004	0.004	0.0005-0.003	0.010-0.025	0.020-0.025	0.008-0.015	0.007-0.014
6.	6500	9500	19,500-23,000	13,000-14,900	20,000-27,000	4500-6000	4000-5500	4300-5000	3400-3560
7.	3.0	1-2	1-4	<1	0.5-3	100-600	200-700	3-8	40-80
8.					11,000	4500-5400	3500-4300	4600	3850
9.	6000	16,000	21,000	11,000-23,000	26,000	5500-8000	3500-8000	7500	3000
10.	9500	14,000	29,000-32,000	17,500-23,000	26,000-34,000	6000-8000	5000-7000	8500-9200	5500-6500
11.	750	480	1100		2500	165-225	100-170	450-575	375-450
12.						150-300			
13.	850	550	1700-1800	1800-2400	2450-2550	170-250	130-200	450-625	360-450
						50	40	400	320
						35	30		
14.	0.6	<0.5	1.4-1.5	0.5-1.0	0.8-1.2	0.4-1.0	1.0-20.0	0.4-0.6	0.6-1.0
15.	R110	R123	R123	R121	R123	R80-102	R50-96	R94-110	R78-97
16.	11	49	22	20	6-8.9	81-100	68-95	55-80	28-40
17.	230	275	485-505	500	500-505	120-140	115-140	180-270	150-165
	250				>505	225-250	185-220	265-290	210-250
18.		6.9	6.9-10.7		17.9	2.8	3.5-4.0	7.6	
19.	1.45	1.3	1.60-1.65	1.8-1.9	1.45-1.47	0.900-0.910	0.890-0.905	1.23-1.27	1.22-1.25
20.	0.03	<0.02	0.02-0.05	0.03	0.01-0.15	0.01-0.03	0.03	0.01-0.03	0.02-0.05
									0.1
21.		380	375-450	340-400		600	600	500	410
	Thermofil	Phillips	RTP; LNP; Phillips; Thermofil; Wilson-Fiberfil	RTP; LNP; Phillips	LNP; Thermofil; Wilson-Fiberfil	Amoco; Bamberger; Eastman; El Paso; Ferro; Monmouth; Schulman; Shuman; USS Chem.; Arco; Exxon; Gulf; Himont USA; Northern Petrochemical; Phillips; Shell; Soltex; Thermofil; Washington Penn	Bamberger; Eastman; El Paso; USS Chem.; * Arco; Gulf; Himont USA; Northern Petrochemical; Phillips; Shell; Soltex; Washington Penn	Amoco; Eastman; Ferro; RTP; Schulman; Arco; Gulf; Himont USA; Polifil; Shell; Thermofil; Washington Penn; Wilson-Fiberfil	Ferro; Schulman; Gulf; Himont USA; Polifil; Thermofil; Washington Penn; Wilson-Fiberfil

(continued)

TABLE 4.4 (continued)

| | | | Polypropylene (Cont'd) | | | Polystyrene and styrene copolymers | | |
| | | | | | | Polystyrene homopolymers | | Styrene copolymers |
Materials	Properties	ASTM test method	40% glass fiber-reinforced homopolymer	EMI shielding (conductive); 30% PAN carbon fiber	Polyallomer	High and medium flow	Heat-resistant	Styrene-acrylonitrile copolymer
Processing	1. Melting temperature, °C T_m (crystalline)		168	168	120-135			
	T_g (amorphous)					88-105	104-110	120
	2. Processing temperature range, °F (C = compression; T = transfer; I = injection; E = extrusion)		I: 500-550	I: 360-470	I: 430-445	C: 300-400 I: 350-500 E: 350-500	C: 300-400 I: 350-500 E: 350-500	C: 300-400 I: 360-550 E: 360-450
	3. Molding pressure range, 10^3 p.s.i.		10-25		1-2	5-20	5-20	5-20
	4. Compression ratio					3	3-5	3
	5. Mold (linear) shrinkage, in./in.	D955	0.003-0.005	0.001-0.003	0.010-0.020	0.004-0.007	0.004-0.007	0.003-0.005
Mechanical	6. Tensile strength at break, p.s.i.	D638[b]	8500-15,000	6800	3000-3800	5200-7500	7100-8200	10,000-11,900
	7. Elongation at break, %	D638[b]	2-4	0.5	400-500	1.2-2.5	2.0-3.6	2-3
	8. Tensile yield strength, p.s.i.	D638[b]			3000-3400		8000-8150	10,400-12,000
	9. Compressive strength (rupture or yield), p.s.i.	D695	8900-9800			12,000-13,000	13,000-14,000	14,000-15,000
	10. Flexural strength (rupture or yield), p.s.i.	D790	10,500-22,000	9000		10,000-14,600	13,000-14,000	11,000-15,000
	11. Tensile modulus, 10^3 p.s.i.	D638[b]	1100-1500	1750		330-475	450-485	475-560
	12. Compressive modulus, 10^3 p.s.i.	D695				480-490	495-500	530-580
	13. Flexural modulus, 10^3 p.s.i. 73°F.	D790	950-1000	1650	70-110	380-490	450-500	500-580
	200°F.	D790						
	250°F.	D790						
	300°F.	D790						
	14. Izod impact, ft.-lb./in. of notch (⅛-in. thick specimen)	D256A	1.4-2.0	1.1	1.7-3.8	0.35-0.45	0.4-0.45	0.4-0.6
	15. Hardness Rockwell	D785	R102-111		R50-85	M60-75	M75-84	M80, R83
	Shore/Barcol	D2240/ D2583						
Thermal	16. Coef. of linear thermal expansion, 10^{-6} in./in./°C.	D696	27-32		83-100	50-83	68-85	65-68
	17. Deflection temperature under flexural load, °F. 264 p.s.i.	D648	300-310	245	124-133	169-202	194-217	214-220
	66 p.s.i.	D648	330		165-192	155-204	200-224	220-224
	18. Thermal conductivity, 10^{-4} cal.-cm./sec.-cm.^{2}.°C.	C177	8.4-8.8		2-4	3.0	3.0	3.0
Physical	19. Specific gravity	D792	1.22-1.23	1.04	0.896-0.899	1.04-1.05	1.04-1.05	1.07-1.08
	20. Water absorption (⅛-in. thick specimen), % 24 hr.	D570	0.05-0.06	0.12	<0.01	0.01-0.03	0.01	0.15-0.25
	Saturation	D570	0.09-0.10			0.01-0.03	0.01	0.5
	21. Dielectric strength (⅛-in. thick specimen), short time, v./mil	D149	500		800	500-575	500-525	425
	SUPPLIERS[a]		Ferro; RTP; Schulman; USS Chem. Adell; Himont USA; LNP; Soltex; Thermofil; Washington Penn; Wilson-Fiberfil	Wilson-Fiberfil	Eastman	Amoco; Bamberger; Cosden; Mobil; Schulman; USS Chem.; Am. Hoechst; Dow Chem.; Gulf; Hammond; Monsanto; Montedison	Amoco; Cosden (see ad, p. 83); USS Chem.; Am. Hoechst; Dow Chem.; Gulf; Hammond; Kama; Monsanto	Dow Chem.; Monsanto; Montedison

[a]See Ref. 10 for suppliers of specialty materials and custom compounds.
[b]Tensile test method varies with material: D638 is standard for thermoplastics; D651 for rigid thermosetting plastics; D412 for elastomeric plastics; D882 for thin plastics sheeting.

TABLE 4.4 (continued)

Polystyrene and styrene copolymers (Cont'd)

	Styrene copolymers (Cont'd)			High-impact polystyrene and high-impact styrene copolymers					
								Flame-retarded, UL-V0	
	High heat-resistant styrene copolymer	Styrene methyl meth-acrylate copolymer	Styrene-maleic anhydride copolymer	Rubber-modified polystyrene homo-polymer	High heat-resistant styrene copolymer	Impact-modified styrene-maleic anhydride polymers	Olefin rubber-modified styrene-acrylonitrile copolymer	Rubber-modified homo-polymer	High heat-resistant styrene copolymer
1.									
		91-97	114	93-105		105			
2.	I: 450-550	I: 375-475	I: 375-475 E: 400-500	I: 350-525 E: 375-500	I: 425-540 E: 400-500	I: 400-510 E: 425-525	I: 450-510 E: 430-510	I: 400-450 E: 375-425	I: 425-525 E: 400-500
3.		5-20	0.6-2	10-20		0.6-17	5-17	6-15	
4.		2.5-3.5		4		2.5	3-4	3	
5.	0.005	0.002-0.006	0.005	0.004-0.007	0.003-0.006	0.004-0.006	0.005-0.007	0.003-0.006	0.005
6.	7100-8100	8100-9700	8100	2700-6200	4600-5800	4500-5500		2650-4100	3000
7.	1.7-1.9	2.1-3.0	1.8	20-65	10-20	1.8-50	50-80	30-50	1.9
8.				2700-6000		4500-6000	5600-6000	3100-4400	
9.	11,400-14,200								
10.		14,500-15,800	14,200	4500-8300	8500-10,500	7600-10,500	7200-8900	5900-7500	6300
11.	440-490	440-500	490	240-370	280-330	270-350	300	240-300	244
12.		440-480							
13.	450-490		470	260-360	320-370	280-360	280-300	320-330	270
14.	0.4-0.6	0.2-0.3	0.6	0.95-3.5	1.5-4.0	3-6	13-15	1.9-3.2	1.3
15.		M72-80	L106	L50-82	L75-95	L75-109; R98	R100-102	L38-65	
16.	65-67	40-72	80		67	47-82	80		
17.	226-249	205-210	226 annealed	170-205	230-248	198-235	208-210 annealed	200-205	230
			165-200					176-181	
18.									
19.	1.07-1.10	1.09-1.13	1.05	1.03-1.06	1.05-1.08	1.05-1.24	1.02	1.16-1.17	1.20
20.	0.1	0.11-0.15		0.05-0.07	0.1	0.23-0.25	0.09	0.0	0.1
			0.1						
21.						415-480	420	550	
	Arco	Richardson	Arco	**Amoco; Bamberger** **Cosden** **Mobil; Shuman** **USS Chem.;** Am. Hoechst; Dow Chem.; Gulf; Hammond; Monsanto; Montedison	Arco; Monsanto; Montedison	Arco; Monsanto	Uniroyal	**Schulman** Am. Hoechst; Montedison	Arco; Montedison

(continued)

TABLE 4.4 (continued)

	Properties	ASTM test method	Polystyrene and styrene copolymers (Cont'd)				Polyurethane	
			Glass fiber-reinforced polystyrene and styrene copolymers				Casting resins	
			20% long and short glass-reinforced homo-polymer	20% long glass-reinforced styrene-acrylonitrile	20% glass reinforced high heat-resistant copolymer	EMI shielding (conductive); 20% PAN carbon fiber	Liquid	Unsaturated
Processing	1. Melting temperature, °C. T_m (crystalline)						Thermoset	Thermoset
	T_g (amorphous)		115	120				
	2. Processing temperature range, °F. (C = compression; T = transfer; I = injection; E = extrusion)		I: 400-460	I: 400-500	I: 425-550	I: 430-500	C: 185-250	
	3. Molding pressure range, 10^3 p.s.i.						0.1-5	
	4. Compression ratio							
	5. Mold (linear) shrinkage, in./in.	D955	0.001-0.003	0.001-0.003	0.003	0.0005-0.003	0.020	
Mechanical	6. Tensile strength at break, p.s.i.	D638[b]	10,000-12,000	15,500-16,000	10,700-12,000	14,000	175-10,000	10,000-11,000
	7. Elongation at break, %	D638[b]	1.3	1.2-1.3	1.4-1.9	1	100-1000	3-6
	8. Tensile yield strength, p.s.i.	D638[b]						
	9. Compressive strength (rupture or yield), p.s.i.	D695	16,000-17,000	17,000-21,000			20,000	
	10. Flexural strength (rupture or yield), p.s.i.	D790	14,000-18,000	20,000-22,700	16,300-19,700	20,700	700-4500	19,000
	11. Tensile modulus, 10^3 p.s.i.	D638[b]	900-1100	1200-1710	850-900	2000	10-100	
	12. Compressive modulus, 10^3 p.s.i.	D695					10-100	
	13. Flexural modulus, 10^3 p.s.i. 73°F.	D790	950-1000	1100-1280	800-870	1900	10-100	610
	200°F.	D790						
	250°F.	D790						
	300°F.	D790						
	14. Izod impact, ft.-lb./in. of notch (⅛-in. thick specimen)	D256A	0.9-2.5	2.5-3.0	2.1-2.6	0.7	25 to flexible	0.4
	15. Hardness Rockwell	D785	M80-95	M89-100				
	Shore/Barcol	D2240/D2583					Shore A10, D90	Barcol 30-35
Thermal	16. Coef. of linear thermal expansion, 10^{-6} in./in./°C.	D696	39.6	23.4-41.4	20		100-200	
	17. Deflection temperature 264 p.s.i. under flexural load, °F.	D648	215-220	210-230	234-236	220	Varies over wide range	190-200
	66 p.s.i.	D648	225-230			230		
	18. Thermal conductivity, 10^{-4} cal.-cm./sec.-cm².-°C.	C177					5	
Physical	19. Specific gravity	D792	1.20	1.22-1.40	1.21-1.22	1.14	1.1-1.5	1.05
	20. Water absorption (⅛-in. thick specimen), % 24 hr.	D570	0.1	0.1-0.2	0.1	0.1	0.2-1.5	0.1-0.2
	Saturation	D570	0.3	0.7				
	21. Dielectric strength (⅛-in. thick specimen), short time, v/mil	D149					300-500	
	SUPPLIERS[a]		RTP, LNP, Thermofil, Wilson-Fiberfil	LNP, Thermofil, Wilson-Fiberfil	Arco, LNP, Wilson-Fiberfil	Wilson-Fiberfil	Union Carbide; Upjohn; Conap; Emerson & Cuming; Hexcel; Hysol; Thermoset Plastics	Dow Chem.; Emerson & Cuming; Hexcel; Hysol

[a]See Ref. 10 for suppliers of specialty materials and custom compounds.
[b]Tensile test method varies with material: D638 is standard for thermoplastics; D651 for rigid thermosetting plastics; D412 for elastomeric plastics; D882 for thin plastics sheeting.

TABLE 4.4 (continued)

#	Polyurethane (Cont'd)		Polyvinylidene chloride copolymers			Silicone		Silicone epoxy
				Barrier film		Casting resins	Molding and encapsulating compounds	
	10-20% glass fiber-reinforced molding compounds	EMI shielding (conductive); 30% PAN carbon fiber	Injection molding	Unplasticized	Plasticized	Flexible (including RTV)	Mineral- and/or glass-filled	Molding and encapsulating compound
1.			172	160	172	Thermoset	Thermoset	Thermoset
	120-160		-15	0-2	-15			
2.	I: 360-410	I: 410-450	C: 260-350 / I: 300-400 / E: 300-400	E: 320-390	E: 340-400		C: 280-360 / I: 330-370 / T: 330-370	C: 350
3.	8-11		5-30	3-30	5-30		0.3-6	0.4-1.0
4.			2.5	2-2.5	2-2.5		2.0-8.0	
5.	0.007-0.010	0.001-0.002	0.005-0.025	0.005-0.025	0.005-0.025	0.0-0.006	0.0-0.005	0.005
6.	4800-6500	13,000	3500-5000	2800	3500	350-1000	4000-6500	6000-8000
7.	3-48	20	160-240	350-400	250-300	100-700	<5	
8.			2800-3800		4900			
9.	5000		2000-2700				10,000-16,000	28,000
10.	5500-6200	9000	4200-6200				8000-14,000	17,000
11.	0.95-140	500	50-80	50-80	50-80			
12.			55-95					
13.	90	500	55-95				1000-2500	
14.	14-No break	10	0.4-1.0	0.3-1.0	0.3-1.0		0.25-8.0	0.3
15.	R45-55		M60-65	R98-106	R98-106		M80-95	
						Shore A15-65		
16.	34		190	190	190	300-800	20-50	30
17.	115-130	180	130-150	130-150	130-150		>500	
	140-145							
18.			3	3	3	3.5-7.5	7.18	16
19.	1.22-1.36	1.33	1.65-1.72	1.65-1.70	1.68-1.72	0.99-1.5	1.80-2.05	1.84
20.	0.4-0.55		0.1	0.1	0.1		0.15	
	1.5						0.15-0.40	
21.	600		400-600		400-600	550	200-425	246
	Union Carbide; LNP; Thermofil; Wilson-Fiberfil	Wilson-Fiberfil	Dow Chem.	Dow Chem.	Dow Chem.	Dow Corning; Emerson & Cuming; GE, Waterford; Thermoset Plastics	Fiberite; Dow Corning; GE, Waterford	Dow Corning

(continued)

TABLE 4.4 (continued)

	Properties	ASTM test method	Polysulfone — Injection molding, flame-retarded, extrusion	Mineral-filled	30% glass fiber-reinforced	EMI shielding (conductive); 30% carbon fiber	Polyaryl-sulfone	Polyethersulfone — Unfilled
Processing	1. Melting temperature, °C. T_m (crystalline)							
	$\quad T_g$ (amorphous)		190	190	189-190		220	230
	2. Processing temperature range, °F. (C = compression; T = transfer; I = injection; E = extrusion)		I: 625-750 E: 600-700	I: 675-775	I: 600-700	I: 600-700	I: 630-800 E: 620-750	I: 590-750
	3. Molding pressure range, 10^3 p.s.i.		5-20	10-20		10-20	5-20	10-20
	4. Compression ratio		2.5-3.5	2.5-3.5		2.5-3.5		
	5. Mold (linear) shrinkage, in./in.	D955	0.007	0.004-0.005	0.001-0.003	0.001	0.007-0.008	0.007
Mechanical	6. Tensile strength at break, p.s.i.	D638[b]		9500-9800	14,500	23,500	9000	
	7. Elongation at break, %	D638[b]	50-100	2-5	1.5	2	60	30-80
	8. Tensile yield strength, p.s.i.	D638[b]	10,200				10,400-12,000	12,200
	9. Compressive strength (rupture or yield), p.s.i.	D695	40,000		19,000	25,000		
	10. Flexural strength (rupture or yield), p.s.i.	D790	15,400	14,300-15,400	20,000	32,000	12,400-16,000	18,650
	11. Tensile modulus, 10^3 p.s.i.	D638[b]	360	550-650	1350	2150	310-385	350
	12. Compressive modulus, 10^3 p.s.i.	D695	374					
	13. Flexural modulus, 10^3 p.s.i. 73° F.	D790	390	600-750	1050	2050	330	375
	200° F.	D790	370	570-710				
	250° F.	D790	350	550-690				
	300° F.	D790	310	510-650				
	14. Izod impact, ft.-lb./in. of notch (⅛-in. thick specimen)	D256A	1.2	0.65-1.0	1.1	1.2	1.6-12	1.6
	15. Hardness Rockwell	D785	M69	M70-74	M90-100	M80		M88
	Shore/Barcol	D2240/ D2583						
Thermal	16. Coef. of linear thermal expansion, 10^{-6} in./in./°C.	D696	56	34-39	25	6	31-49	55
	17. Deflection temperature under flexural load, °F. 264 p.s.i.	D648	345	345-354	350	365	400	397
	66 p.s.i.	D648	358		360	380		
	18. Thermal conductivity, 10^{-4} cal.-cm./ sec.-cm.^{2}.°C.	C177	6.2					3.2-4.4
Physical	19. Specific gravity	D792	1.24-1.25	1.48-1.61	1.46	1.36	1.29-1.37	1.37
	20. Water absorption (⅛-in. thick specimen), % 24 hr.	D570	0.3		0.3	0.15		0.43
	Saturation	D570	0.7	0.5-0.6			1.1-1.85	
	21. Dielectric strength (⅛-in. thick specimen), short time, v./mil	D149	425	450			370-380	400
SUPPLIERS[a]			Union Carbide	Union Carbide	RTP, LNP; Thermofil; Wilson-Fiberfil	Thermofil; Wilson-Fiberfil	Union Carbide	ICI Americas

[a]See Ref. 10 for suppliers of specialty materials and custom compounds.
[b]Tensile test method varies with material: D 638 is standard for thermoplastics; D 651 for rigid thermosetting plastics; D 412 for elastomeric plastics; D 882 for thin plastics sheeting.

TABLE 4.4 (continued)

Sulfone polymers (Cont'd)					Thermoplastic elastomer			
Polyethersulfone (Cont'd)		Modified polysulfone			Polyolefin			Block copolymers of styrene and butadiene or styrene and isoprene
20% glass fiber-reinforced	EMI shielding (conductive); 30% carbon fiber	Injection molding	Mineral-filled	30% glass fiber-reinforced	Low and medium hardness	High hardness	Polyester	
1.							145-217	
225					165	165		
2. I: 630-720	I: 680-720	I: 540-590	I: 575-650	I: 520-600	C: 350-450 I: 360-475 E: 380-450	C: 370-450 I: 390-480 E: 410-475	I: 340-480 E: 340-480	C: 250-325 I: 300-425 E: 370-400
3. 10-20	10-20	10-20	5-20	5-20	4-19	6-10	6-15	3-30
4. 2.0-3.5	2.5-3.5	2	2.5-3.5	2	2-3.5	2-3.5	3-3.5	2.0-2.5
5. 0.003-0.005	0.0005-0.002	0.006-0.007	0.006-0.007	0.001-0.003	0.015-0.020	0.015-0.017	0.003-0.014	0.001-0.005
6. 17,400-18,000	26,000-30,000	6500		15,000-19,000	650-2300	2750-4000	2000-6400	600-4350
7. 2-3.5	1.7-2.5	30	50-100	2.5-3.0	150-530	600	170-550	80-1350
8.		7400	10,500			1600	1350-3900	3700
9. 19,500	22,000							
10. 24,500	36,000-38,000	13,300	16,500	23,000-25,500			5100	
11. 856-860	2120-2900	346	400	830-1000	1.1-16.4	18.1-34.1	7.5-72.5	0.8-235
12.							7.5-48	3.6-120
13. 850-856	2000-2500	365	480	900-1240	1.5-14.9	20.4-50.3	8-85	4-150
850			370					
840			340					
840			320					
14. 1.4-1.5	1.2-1.6	9.5	1.1	1.0-2.0	No break		3.9-No break	No break
15. M98-99	R123	R117	M74	M80				
					Shore A64-92	Shore D40-50	Shore D40-72	Shore A40-D65
16. 26	10	65	53	27		90-190		67-137
17. 410-426	415-420	302	325	320-330				<0-170
420	420-425	320		347				<0-190
18.		7.1			4.5-5.0		3.6-4.5	3.6
19. 1.51	1.47-1.48	1.13	1.30	1.47	0.88-0.98	0.94-0.95	1.16-1.25	0.90-1.2
20. 0.35	0.29	0.25		0.14-0.20	0.01		0.3-1.6	0.09-0.39
1.65			0.8	0.43				
21. 420		435	460	400	410-445	450-465	400-460	300-520
ICI Americas LNP; Thermofil; Wilson-Fiberfil	Thermofil; Wilson-Fiberfil	Union Carbide	Union Carbide	Union Carbide; Wilson-Fiberfil	Du Pont; Schulman Union Carbide; Uniroyal; Exxon; Monsanto; Montedison	Monsanto	Du Pont; GAF	Phillips; Shell

(continued)

TABLE 4.4 (continued)

	Properties	ASTM test method	Block copolymers of styrene and ethylene or butylene	Solution coating resins Polyester	Solution coating resins Polyether	Polyurethane Molding and extrusion Polyester Low and medium hardness	Polyester High hardness	Polyether Low and medium hardness	Polyether High hardness
Processing	1. Melting temperature, °C T_m (crystalline)								
	T_g (amorphous)			−20 to +16	−49	120-160	120-160	120-160	120-160
	2. Processing temperature range, °F (C = compression; T = transfer; I = injection; E = extrusion)		C: 300-380 I: 350-480 E: 330-380			I: 380-435 E: 370-410	I: 410-440 E: 370-410	I: 350-430 E: 340-410	I: 400-435 E: 380-440
	3. Molding pressure range, 10^3 p.s.i.		1.5-20			0.8-1.4	0.8-1.4	0.6-1.2	1-1.4
	4. Compression ratio		2.5-5.0						
	5. Mold (linear) shrinkage, in./in.	D955	0.006-0.022			0.008-0.015	0.005-0.015	0.008-0.015	0.008-0.012
Mechanical	6. Tensile strength at break, p.s.i.	D638[b]	1000-3000	4500-7900	5500	3300-8400	4000-11,000	1500-6750	6000-7240
	7. Elongation at break, %	D638[b]	600-850	290-630	530	410-620	110-550	475-1000	340-425
	8. Tensile yield strength, p.s.i.	D638[b]							
	9. Compressive strength (rupture or yield), p.s.i.	D695							
	10. Flexural strength (rupture or yield), p.s.i.	D790							
	11. Tensile modulus, 10^3 p.s.i.	D638[b]		0.33-1.45[c]	0.7[c]				
	12. Compressive modulus, 10^3 p.s.i.	D695							
	13. Flexural modulus, 10^3 p.s.i. 73° F.	D790	4-100						
	200° F.	D790							
	250° F.	D790							
	300° F.	D790							
	14. Izod impact, ft.-lb./in. of notch (⅛-in. thick specimen)	D256A	No break						
	15. Hardness Rockwell	D785							
	Shore/Barcol	D2240/ D2583	Shore A50-90	Shore A70-D54		Shore A55-95	Shore D46-78	Shore A70-92	Shore D55-75
Thermal	16. Coef. of linear thermal expansion, 10^{-6} in./in./°C.	D696							
	17. Deflection temperature under flexural load, °F 264 p.s.i.	D648							
	66 p.s.i.	D648							
	18. Thermal conductivity, 10^{-4} cal.-cm./sec.-cm.^{2}.°C.	C177							
Physical	19. Specific gravity	D792	0.9-1.2	1.19-1.22	1.11	1.17-1.25	1.15-1.28	1.10-1.20	1.14-1.21
	20. Water absorption (⅛-in. thick specimen), % 24 hr.	D570	0.17-0.42				0.3		
	Saturation	D570							
	21. Dielectric strength (⅛-in. thick specimen), short time, v./mil	D149						470	470
SUPPLIERS[a]			Concept Polymer; Dow Chem.	Goodrich	Goodrich	Upjohn; Dainippon; Goodrich; Mobay; Ohio Rubber	Upjohn; Goodrich; Mobay; Ohio Rubber	Upjohn; Goodrich; Mobay; Ohio Rubber	Upjohn; Goodrich; Mobay; Ohio Rubber

[a]See Ref. 10 for suppliers of specialty materials and custom compounds.
[b]Tensile test method varies with material: D638 is standard for thermoplastics; D651 for rigid thermosetting plastics; D412 for elastomeric plastics; D882 for thin plastics sheeting.
[c]Secant modulus at 100% elongation.
Source: Ref. 10.

TABLE 4.4 (continued)

	Urea		Vinyl polymers and copolymers						
			Polyvinyl chloride and polyvinyl chloride-acetate molding compounds, sheets, rods, and tubes			Molding and extrusion compounds			
	Alpha cellulose-filled	PVC molding compound, 15% glass fiber-reinforced	Rigid	Flexible, unfilled	Flexible, filled	Vinyl formal	Chlorinated polyvinyl chloride	Vinyl butyral, flexible	PVC/acrylic blends
1.	Thermoset								
		75-105	75-105	75-105	75-105	105	110	49	
2.	C: 275-350 I: 290-320 T: 270-300	I: 270-405	C: 285-400 I: 300-415	C: 285-350 I: 320-385	C: 285-350 I: 320-385	C: 300-350 I: 300-400	C: 350-400 I: 395-440 E: 360-415	C: 280-320 I: 250-340	I: 360-390 E: 390-410
3.	2-20	8-25	10-40	8-25	1-2	10-30	15-40	0.5-3	2-3
4.	2.2-3.0	1.6-2.2	2.0-2.3	2.0-2.3	2.0-2.3		1.5-2.5		2-2.5
5.	0.006-0.014	0.001	0.002-0.006	0.010-0.050	0.008-0.035 0.002-0.008(trans.)	0.001-0.003	0.003-0.007		0.003
6.	5500-13,000	9500	5900-7500	1500-3500	1000-3500	10,000-12,000	6800-9000	500-3000	6400-7000
7.	<1	2-3	40-80	200-450	200-400	5-20	4-65	150-450	35-100
8.			5900-6500				6000-8000		
9.	25,000-45,000	9000	8000-13,000	900-1700	1000-1800		9000-22,000		6800-8500
10	10,000-18,000	13,500	10,000-16,000			17,000-18,000	14,500-17,000		10,300-11,000
11.	1000-1500	870	350-600			350-600	341-475		340-370
12.							335-600		
13.	1300-1600	750	300-500				380-450		350-380
14.	0.25-0.40	1.0	0.4-22	Varies over wide range	Varies over wide range	0.8-1.4	1.0-5.6	Varies over wide range	1-12
15.	M110-120	R118				M85	R117-122	A10-100	R106-110
			Shore D65-85	Shore A50-100	Shore A50-100				
16.	22-36		50-100	70-250		64	68-76		68-79
17.	260-290	155	140-170			150-170	202-234		167-185
		165	135-180				215-247		172-189
18.	7-10		3.5-5.0	3-4	3-4	3.7	3.3		
19.	1.47-1.52	1.54	1.30-1.58	1.16-1.35	1.3-1.7	1.2-1.4	1.49-1.58	1.05	1.26-1.35
20.	0.4-0.8	0.01	0.04-0.4	0.15-0.75	0.5-1.0	0.5-3.0	0.02-0.15	1.0-2.0	0.09-0.16
21.	300-400	600-800	350-500	300-400	250-300	490		350	480
	Am. Cyanamid; Budd; MKB; Patent Plastics; Perstorp	LNP; Thermofil	Alpha Chem. **Occidental;** **Union Carbide;** Air Products; Borden; Colorite; Conoco; Georgia-Pac.; Goodrich; Goodyear; Keysor-Century; Novatec; Pantasote; Stauffer; Tenneco	Alpha Chem. **Occidental;** **Schulman;** Air Products; Borden; Colorite; Conoco; Georgia-Pac.; Goodrich; Keysor-Century; Pantasote; Tenneco	Alpha Chem. **Occidental;** **Union Carbide;** Air Products; Borden; Colorite; Conoco; Georgia-Pac.; Goodrich; Keysor-Century; Pantasote; Stauffer; Tenneco	Monsanto	Goodrich	**Union Carbide;** Monsanto	Sumitomo

New Materials

With all plastics materials, their usage in the gear application must be
verified by testing. All new plastics materials considered for gearing
must undergo the same rigorous manufacturing scrutiny and application testing.

As discussed in relation to shrinkage and warpage in Chapter 3,
new materials are constantly under development and all cannot be included here nor can they be ignored. Also, a vast array of filled materials are available and data are presented in tables that give wear
factors, material data, and coefficients of friction of plastics against
plastics and plastics against steel (9). "Super-tough" materials have
been developed that may apply to gearing where impact is a problem
and a detriment to long life. DuPont provides data with comparative
tables as a guideline for material selection using Delrin T (toughened
resin), designated Delrin 500T, Delrin ST (super-tough) designated
Delrin 100ST or Zytel ST, designated Zytel ST801 or ST811. Delrin
ST and Delrin T are acetal-elastomer alloys that retain all the bene-
ficial properties of standard Delrin acetal resins (16,17,18). With
these improvements and all other mechanical properties remaining un-
changed, the material seems to have good potential for gearing applications but must be tested in candidate applications.

Another DuPont development is a weatherable black grade of unreinforced 6/6 nylon called Zytel 105F. Manufacturing productivity
for small, difficult-to-mold parts has been reportedly increased by
50% for those that require ultraviolet resistance.

Other materials are available that include an injection-molding grade
of polypropylene for control of electrostatic discharge (ESD). The material is designated by LNP Corporation as Stat-Kon M-1® (19). This
material series is not engineered for EMI attenuation.

Reinforcements, Fillers, and Lubricants

In the literature, reinforcements, fillers, and lubricants are sometimes
broadly referred to as fillers because they are materials added to the
basic plastic material. Strictly speaking, each is added for a specific
purpose. Reinforcements are usually thought of as an added material
whose purpose is to alter one or more mechanical properties of the basic material. Fillers are usually regarded as materials that fill the
bulk of the molded part and are of a lower cost than the basic plastics resin. Lubricants are added for the specific purpose of changing
the lubricity at the surface of the material or for dissipation of heat.
Internally filled plastics provide lubrication either by migration of the
lubricant or by release due to surface thermal change or by wear.

Glass is a plastics gear reinforcement often used to alter both long- and short-term mechanical properties of the basic material resin. Glass fibers are preferred over unsized milled glass or glass beads. Although glass fibers improve creep resistance, thermal conductivity, and heat-deflection temperature, wear of the mating gear and an increased coefficient of friction are disadvantages. For this reason PTFE is usually recommended when glass is added to the base resin. When glass is added, the addition of silicone alone is not recommended; however, PTFE can be used without the addition of silicone.

As the cost of petroleum-based resins rise, mineral fillers play an increasingly important role. Advantages are

Improved moldability
Upgrading of stability
General cost reduction
Mineral filler costs that are more stable than base resin costs

Fillers are often used to enhance or provide properties such as magnetism (barium ferrite), conductivity (graphite), and lubricity (PTFE). Other filler materials may be used. Since fillers reduce mechanical properties, they are often used with other fibrous reinforcements that offset the reduced mechanical properties.

Use of mineral fillers provides physical and thermal property improvements by the nature of the mineral, particle size and shape, surface area, size distribution, and amount of filler. Past improvements include increase of density and heat-deflection temperature, and reduction of thermal expansion, mold cycle times, and post-mold shrinkage. Some mineral fillers impart flame-retardant characteristics, increase strength, and are classed as reinforcements. Common mineral fillers are calcium carbonate, talc, silica, wollastonite, clay, Franklin fiber (calcium sulfate fiber), mica, glass beads, and alumina trihydrate. A profile of filler characteristics is given in Table 4.5, and properties of mineral-filled nylon 6/6 resin are presented in Table 4.6 (20). Theberge states that the effects of these minerals can be expected to be similar to those of other crystalline resins, such as the thermoplastic polyesters and other grades of nylon. The results would not be generally the same for amorphous resins such as the polycarbonates and polyolefins or with other relatively low-polarity resins. Creep data due to loading and addition of reinforcements and fillers are provided by Newby and Theberge (21). If extensive service is required of the gears, this information and the latest literature should be checked for recent developments in this important area.

TABLE 4.5 Profile of Filler Characteristics

Shape

The shape of filler particles determines mechanical properties of a composite as much as or more than the composition of the material itself. Fibers and prismatic shapes usually increase both tensile and flex strength in all three orthogonal directions; spheres and cubes, having an $L/D = 1$, raise neither property. Between these shapes are flake, or plate-like materials, that can reinforce in two directions because they usually become oriented in the direction of flow during molding. All shapes increase modulus as well as contribute to dimensional and thermal stability of a composite.

Size

Properties affected by particle size are primarily tensile, flexural, and impact strength. Particles range in mean size from about 1 to 10 microns for the natural minerals and to 30 microns for glass beads. Large particles cause greater stress concentrations in the matrix than small ones. In general, small particle sizes produce better strength properties (or less strength reduction) than do large ones.

Size distribution

Variation in particle size for most minerals ranges from about 20:1 to 40:1. However, any given lot of clay or talc can have size distributions as great as 1,000:1. In general, the smaller and more uniform the particle size, the better the strength properties of a composite. Even a few large particles in an otherwise uniform distribution of small particles can reduce strength significantly.

Surface area

Among the mineral fillers, clay has the highest surface area (to 20 m^2/g) and silica the lowest (1 m^2/g). Effects of fillers on viscosity control during the molding of composites depend both on surface area available and on the degree of wetting-out of that surface by a polymer or coupling agent. A well-wetted mineral filler flows more easily and helps provide better mechanical properties in a composite. However, if the surface area is so large that too little resin is available to wet-out the entire area, a composite with poorly dispersed filler and nonuniform properties can result.

Sphere

Cube

Flake or Plate

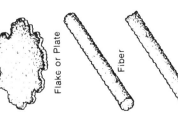

Fiber

Prism

Properties of Commonly Used Mineral Fillers

	Mica	Calcium carbonate	Wollastonite	Glass beads	Alumina trihydrate	Talc	Silica	Franklin fiber	Clay
Water content (%)	<5.0	<2.0	0.5	<0.1	34.6	4.8	<0.1	<1.0	<0.5
Specific gravity	2.74-2.95	2.60-2.75	2.9	2.48	2.42	2.7-2.8	2.65	3.0	2.50
Hardness (Mohs)	2.4-3.0	3.0	4.5	5.5	2.5-3.5	1.0	7.0	2.0	4.0-6.0
Melt point (°C)	1,300[a]	900[a]	1,540	1,200	200-600[b]	Stable to 380	Stable to 573	-	1,810
Color	Gold-brown	White	White	Clear	White	Gray-white	White	White	White
Shape	Plate-like	Prismatic	Fibrous	Spherical	Plate-like	Plate-like	Spherical	Fibrous	Plate-like
Resistance to acids/alkalis	G/G	P/F	P/F	G/P	G/G	G/G	E/P	G/G	G/G

[a]Decomposes.
[b]Loses 34.6% water.
Source: Ref. 20.

TABLE 4.6 Properties of Mineral-Filled Nylon 6/6[a] Resin Composites

	None	Mica	Calcium carbonate	Wollastonite	Glass beads
Specific gravity	1.14	1.50	1.48	1.51	1.46
Mold shrinkage (in./in.)	0.018	0.003	0.012	0.009	0.011
Water absorption, 24 h (%)	1.5	0.6	0.5	0.5	0.6
Tensile strength (psi)	11,800	15,260	10,480	10,480	9,780
Tensile elongation (%)	60	2.7	2.9	3.0	3.2
Flexural strength (10^3 psi)	15	26.0	16.5	17.7	15.8
Flexural modulus (10^3 psi)	410	1,540	660	790	615
Impact strength, Izod (ft-lb/in.)					
Notched	0.06	0.6	0.5	0.6	0.4
Unnotched	No brk	8.1	9.6	9.4	5.5
Heat-deflection temp (°F)	170	460	390	430	410
Thermal expansion (10^{-5} in./in.-°F)	4.5	2.2	2.8	2.3	2.8
Thermal conductivity (Btu-ft^2/in.-h-°F)	1.7	3.9	3.9	4.9	3.4
Oxygen index (% O_2)	27	31	27	27	27
Wear factor (in.^{3}-min/ ft-lb-h)	200	230	180	715	460

Filler[b]						
Aluminum trihydrate	Clay	Talc	Silica	Franklin fiber	Mica (25) glass (15)	Mica (25)[a] glass (15)
1.45	1.47	1.49	1.48	1.52	1.49	1.49
0.008	0.004	0.008	0.005	0.005	0.006	0.0055
0.7	0.4	0.4	0.4	0.7	-	-
9,200	10,850	8,980	13,040	12,240	20,500	20,400
2.8	2.5	2.0	6.4	2.9	3.4	3.8
14.7	23.7	15.3	21.2	18.6	28.5	36.6
645	1,010	925	765	865	1,450	1,645
0.5	0.3	0.6	0.6	0.5	0.8	0.9
6.4	12.3	5.1	31.0	9.3	15.0	11.5
395	390	445	400	435	490	395
3.0	2.6	2.0	2.8	2.8	2.1	2.0
4.9	3.9	4.1	5.3	4.7	3.7	3.6
36	28	34	26	26	-	-
560	1,250	3,150	240	270	-	-

[a]Data are on composites based on nylon 6/6 except for the last column; these data are based on nylon 6.
[b]All composites tested contained 40% (wt) filler and 60% resin, except those in the last two columns which contained 25% mica, 15% glass fibers, and 60% resin.
Source: Ref. 20.

Lubricants used internally in plastics gears are primarily polytetrafluoroethylene (PTFE), silicone, carbon fibers, and graphite powder. The influence of each can be summarized as follows (22):

PTFE: greatly improves surface wear characteristics, reduces coefficient of friction and shock by providing a lubricous film on the mating gear, and minimizes fatigue failure.

Silicone: is partially compatible with base resins; provides boundary or mixed-film lubrication by migration and random molecular movement.

Carbon fibers: has the highest strength, modulus, heat-deflection temperature, creep, and fatigue endurance of commercially available composites. It also has increased thermal conductivity, low coefficient of friction, higher cost, and reduced wear rate compared to glass fibers. It is compatible with other internal lubricants, has low volume resistivity and surface resistivity, and can dissipate static charge if the fill is over 15%.

Graphite powder: has low friction, provides boundary lubrication, with coefficient of friction and wear factors between the base resin and PTFE/silicone versions, is adequate in an aqueous environment, and can withstand high temperatures.

Still other materials have also been used and more are being developed. Designers should not limit themselves if a particular application's demands cannot be met by the few popular materials. Selection of a material depends not only on its material properties but also on the manufacturing method used, part shape and size, molding or machining characteristics, shrinkage rates, moisture absorption, and processing variables. Time spent in consulting with the material supplier or processor can pay dividends in eliminating costly mold reworking, delays in receiving parts, possibility of failures, additional or unnecessary costs, and other potential problems. In view of the dynamic nature of the plastics industry, designers should be satisfied that they have the latest possible material data. Improvement in materials, the development of engineering data, and advancements in mold design and molding technology are constantly offering new possibilities.

References

1. E. Miller, Ed., *Plastics Products Design Handbook, Part A, Materials and Components*, Marcel Dekker, NY (1981).

2. F. R. Estabrook, *Why Not Try Plastic Gears?*, Penton/IPC, Cleveland, OH (April 1982).

3. *Machine Design*, Penton/IPC, Cleveland, OH (April 14, 1983 and June 30, 1983).

4. *The Celcon® Acetal Copolymer Design Manual*, Celanese Plastics Company, Newark, NJ.

5. *Design and Production of Gears in Celcon® Acetal Copolymer*, Celanese Plastics Company, Newark, NJ (1969).

6. *DuPont Delrin® Acetal Resin Design Handbook*, E. I. DuPont DeNemours and Co., Wilmington, DE (1981).

7. *DuPont Zytel® Nylon Resin Design Handbook*, Bulletin E-44971, E. I. DuPont DeNemours and Co., Wilmington, DE.

8. H. R. Clauser, Ed., *Encyclopedia/Handbook of Materials, Parts, and Finishes*, Technomic, Westport, CT (1976).

9. *Fortified Polymers, LNP Engineering Plastics*, Red Book, LNP Corporation, Malvern, PA (1984).

10. *Modern Plastics Encyclopedia*, McGraw-Hill, NY (1984).

11. *The International Plastics Selector, Inc.*, San Diego, CA (1983).

12. *Materials Engineering*, Reinhold, Stamford, CT (monthly).

13. "Mechanical Drives Reference Issue," *Machine Design*, Penton/IPC, Cleveland, OH (1983).

14. *Modern Plastics Encyclopedia*, McGraw-Hill, NY (annual).

15. *Hytrel® Polyester Elastomer Design Handbook*, E. I. DuPont DeNemours and Co., Wilmington, DE (1982).

16. "Toughest Metal-Like Engineering Plastic," *Machine Design News Trends*, Penton/IPC, Cleveland, OH (June 1983).

17. "New Unreinforced Nylon Improves Productivity," *Machine Design News Trends*, Penton/IPC, Cleveland, OH (1983).

18. *Designing With Plastics . . . We Have the Answers!*, Bulletin E-59937, E. I. DuPont DeNemours and Co., Wilmington, DE.

19. *Injection Molding Grade of Polypropylene*, LNP Corporation, Malvern, PA (1982).

20. J. E. Theberge, "How Mineral Fillers Improve Plastics," *Machine Design*, Penton/IPC, Cleveland, OH (June 1980).

21. G. B. Newby, J. E. Theberge, "Long-Term Behavior of Reinforced Thermoplastics." *Machine Design*. Penton/IPC, Cleveland, OH (March 1984).

22. "Making Resins Stronger and Slipperier," *Machine Design Materials Reference Issue*, Penton/IPC, Cleveland, OH (April 1983).

5
Tooth Elements

Nothing has as yet been uncovered about the behavior of plastics in gearing that would indicate any necessity to depart from the principles of design established for gears of other materials. But the physical characteristics of plastics make it essential to adhere to these principles more rigidly than would necessarily be the case in designing gears to be machined of the metals. The design of all plastics gearing should receive the close study normally reserved for metal gears in critical applications (1).

Using these well-established principles, two important topics relating to tooth elements will be clarified. One concerns some questionable concepts, and the second relates to tooth modifications sometimes not considered in the normal practices of general plastics gearing.

Accuracy Requirements

When a gear application has a need for a certain quality class, it is essential that the class be held in the gearing regardless of the material used for the gearing. It is unreasonable to expect that incorrect tolerances, poor profile contact, or other inaccuracies can be overlooked by relying on run-in wear, tooth deflections, and tooth compliance conditions to compensate for poor gearing. The substitution of a Q5 quality class gear for a Q6 quality class gear allows additional inaccuracies in the gear system. For example, a 32-diametral pitch, 1.0-inch-pitch diameter, Q5 quality class gear is permitted more tolerance on tooth-to-tooth and composite action than

the same size gear manufactured to Q6 class tolerances. For Q5, values in ten-thousandths of an inch for tooth-to-tooth and composite
tolerances are 30.1 and 61.5. For Q6, values in ten-thousandths of
an inch for tooth-to-tooth and composite tolerances are 21.5 and 44.0.
It can be seen that the tolerance difference between Q5 and Q6 is significant, particularly in gears of the size used in the example, and
yet this size gear is commonly molded using plastics materials.

The best design method is to calculate the necessary quality class
for the application and specify that class. It serves no useful purpose, however, to specify a gear to a higher classification than is
absolutely needed in the application. A realistic evaluation should
be made remembering that unreasonable quality specifications add to
the parts cost by increasing tooling costs, cycle time, and inspection
effort without adding to the machine function or improving the application.

Tooth Modification

The second topic of tooth modification should be handled routinely in
plastic gearing; however, this is not always the case. A good practice is to adopt the full round radius in the root of the gear tooth.
Since a tooth is essentially a short cantilever beam with the maximum
stress occurring at the fillet of the base of the tooth, the tension side
of the tooth must be considered in relation to high stress and tooth
failures. A stress concentration factor should be used if teeth are
highly loaded. The following provides a calculation for the stress
concentration factor K_t, as presented by Peterson (2) and verified
by several investigators. However, if the gear is designed and manufactured with a full-round root radius, a stress concentration factor
is rarely, if ever, needed.

Peterson's stress concentration factor K_t, for the fillet on the tension side is, for a 14 1/2° pressure angle

$$K_t = 0.22 + \frac{1.0}{\left(\frac{r_f}{t}\right)^{0.2}\left(\frac{h}{t}\right)^{0.4}}$$ (5.1)

and for a 20° pressure angle

$$K_t = 0.18 + \frac{1.0}{\left(\frac{r_f}{t}\right)^{0.15}\left(\frac{h}{t}\right)^{0.45}}$$ (5.2)

Figure 5.1 was constructed from the empirical work of investigators (2) by providing a semicircular fillet radius with a form grinder,

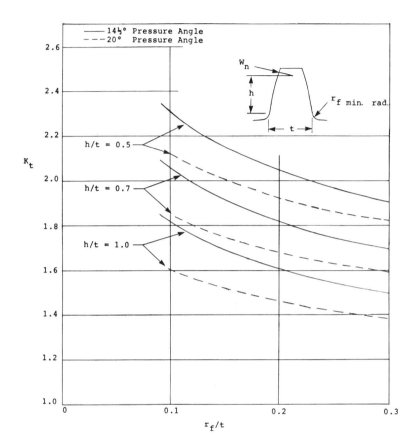

FIGURE 5.1 Stress concentration factor K_t, for a gear tooth fillet on the tension side. (From Ref. 2.)

neglecting load sharing. Figure 5.2 gives the gear notation used. Peterson concluded that a semicircular fillet does not result in a very large decrease in K_t, although the decrease represents a definite available gain. Recent testing of plastics gears with load-sharing and application improvements attests to the fact that substantial benefits can be gained by specifying the full radius at the tooth root (1,3) and are especially necessary for plastic power gearing. Heavily loaded metal gears are routinely modified by providing tip relief to reduce excessive tooth deflection problems. Further information on derating factors is given in Chapter 6.

In plastics gearing, with more than one pair of teeth in mesh, large tooth deflections, high-speed application, or mismatch of mating gear

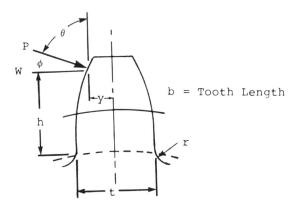

FIGURE 5.2 Gear notation. (From Ref. 2.)

profiles, unusual and often unexpected interference, noise, sliding, and overheating can occur. Many of these undesirable features can be prevented or minimized by relieving the top corners of the teeth, as shown in Figure 5.3 (4). Modification techniques such as controlling the pressure angle, increasing the recess action using long- and short-addendum gears, and relieving the involute are commonly used.

Tooth Systems

Several tooth parameters have been standardized to provide systems that offer interchangeability, common tooling leading to reduced tool and parts inventories, few design calculations, and wide acceptance of the best fundamental design criteria and manufacturing procedures. All of the various recognized systems employ the basic rack (the tooth form if it were cut in a rack or a gear having an infinite number of teeth) to specify pressure angle and tooth proportions. Gears with the same basic proportions are therefore conjugate to the rack and will conform to gears of the same basic pressure angle and data. Table 5.1 lists basic proportions of several systems and Figures 5.4 and 5.5 are the basic fine-pitch and coarse-pitch racks (5,6,7). The only difference is the full round radius in the tooth root of the coarse-pitch system when compared to the fine-pitch system. To compensate, formulas for dedendum, whole depth, clearance, and basic circular tooth thickness on the pitch line also differ, as seen in Table 5.1, columns 4 and 6. Table 5.2 provides a list of standard pitches and the preferred modules for the International System (SI).

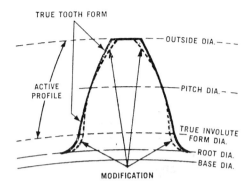

TRUE TOOTH FORM

ACTIVE PROFILE

OUTSIDE DIA.

PITCH DIA.

TRUE INVOLUTE FORM DIA.

ROOT DIA.

BASE DIA.

MODIFICATION

FIGURE 5.3 Modification of tooth profile. (From Ref. 4.)

Involute Profile Generation

An understanding of the geometry of the involute curve can be obtained by studying the many gearing texts in the reference list. In-depth studies were undertaken by Buckingham, Candee, Merritt, Tuplin, and others (8,9,10,11).

The classical explanation can be demonstrated by wrapping a string around a fixed cylinder or disk and tying a pencil on the loose end. The pencil point will inscribe an involute curve if the string is kept tight and unwound from the cylinder (Figure 5.6). The diameter of the disk is equivalent to the base circle in gearing. In Figure 5.7 we find, with A any point on the involute curve,

$$AB = \text{arc subtended by angle } \beta \qquad (5.3)$$

$$AB = OB \times \beta \text{ radians} \qquad (5.4)$$

$$\beta \text{ radians} = AB/OB = \tan \phi \qquad (5.5)$$

$$\phi = \text{pressure angle}$$

Further

$$\Theta = \beta - \phi \qquad (5.6)$$

$$\Theta \text{ radians} = \tan \phi - \phi \text{ radians} \qquad (5.7)$$

Tan ϕ-ϕ radians is by definition the involute of ϕ degrees. For example, to determine the involute of various pressure angles, the

TABLE 5.1 Tooth Proportions of Basic Rack for Standard Involute Gear Systems

		Tooth Proportions	
		1	2
Tooth parameter (of basic rack)	Symbol in Figure 5.5	14 1/2° Full-depth involute system	20° Full-depth involute system
1. System sponsors	-	ASA and AGMA	ASA
2. Pressure angle	ϕ	14 1/2	20°
3. Addendum	a	$1/P_d{}^a$	$1/P_d$
4. Minimum dedendum	b	$1.157/P_d$	$1.157/P_d$
5. Minimum whole depth	h_t	$2.157/P_d$	$2.157/P_d$
6. Working depth	h_k	$2/P_d$	$2/P_d$
7. Minimum clearance	c	$0.157/P_d$	$0.157/P_d$
8. Basic circular tooth thickness on pitch line	t	$1.5708/P_d$	$1.5708/P_d$
9. Fillet radius in basic rack	r_f	1 1/3 X clearance	1 1/2 X clearance
10. Diametral pitch range	-	Not specified	Not specified
11. Governing stand- ard: ASA and AGMA	- -	B6.1 201.02A	B6.1

[a]Diametral pitch.
Source: Ref. 5.

number of radians in the degrees of pressure angle are subtracted from the tangent of the pressure angle, as illustrated in Table 5.3.

Mathematical tables are available for quick determination of the involute function, or it can be calculated as explained above. A table of cosine, involute, and tangent functions is provided for certain

for Various Standard Systems			
3	4	5	6
20° Stub involute system	20° Coarse-pitch involute spur gears	25° Coarse-pitch involute spur gears	20° Fine-pitch involute system
ASA and AGMA	AGMA	AGMA	ASA and AGMA
20°	20°	25°	20°
$0.8/P_d$	$1.000/P_d$	$1.000/P_d$	$1.000/P_d$
$1/P_d$	$1.250/P_d$	$1.250/P_d$	$1.200/P_d + 0.222$ in.
$1.8/P_d$	$2.250/P_d$	$2.250/P_d$	$2.200/P_d + 0.002$ in.
$1.6/P_d$	$2.000/P_d$	$2.000/P_d$	$2.000/P_d$
$0.200/P_d$	$0.250/P_d$	$0.250/P_d$	$0.200/P_d + 0.002$ in.
$1.5708/P_d$	$\pi/2P_d$	$\pi/2P_d$	$1.5708/P_d$
Not standardized	$0.300/P_d$	$0.300/P_d$	Not standardized
Not specified	19.99 and coarser	19.99 and coarser	20 and finer
B6.1	-	-	B6.7
201.02A	201.02	201.02	207.04

selected angles in Appendix A. Intermediate values may be found by interpolation.

Based on the development of the involute function and the relationships of Figures 5.6 and 5.7, we can see that OB = OA X cos ϕ. Point A is the pitch point and OA is the pitch radius. We

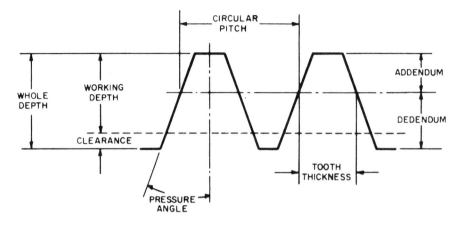

FIGURE 5.4 Basic rack (normal plane), fine-pitch. (From Ref. 6.)

can now relate the base circle diameter to the pitch circle diameter
as

$$D_b = D \cos \phi \qquad\qquad (5.8)$$

It should be clear that the involute curve is cam-shaped and operates
as a cam. By placing two base circles with involute curves together,
as shown in Figure 5.8, the cam surfaces contact and circle O when
rotated will cause angular displacement of circle O_1. The locus of
the points of contact of BB_1 is the common tangent to the two base
circles and is called the line of action. As circle O rotates at a uniform rate of speed in the direction indicated by the arrow, AB length-
ens and AB_1 shortens at the same uniform rate of speed. This then
causes circle O_1 to rotate at a uniform rate of speed. In gearing,
there are uniformly spaced involute curves around the base circles
that provide the cam contact surfaces so that one set of curves are
in contact at all times.

Base circle O is one-half the diameter of base circle O_1. Although
they rotate at the same uniform rate, the larger base circle turns
through one-half the angle of the involute of the smaller base circle.
We then can conclude that mating involute gears are in inverse pro-
portion to the diameters of their base circles and also to the number
of teeth in the gears. The gear ratio then is this same inverse pro-
portion determined by numbers of teeth or base circle diameters or
by input and output relationships.

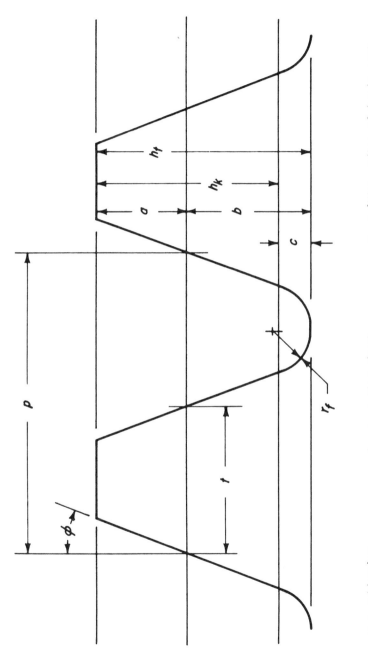

a = addendum
b = dedendum
c = clearance

h_k = working depth
h_t = whole depth
p = circular pitch

r_f = fillet radius of basic rack
t = circular tooth thickness – basic
ϕ = pressure angle

FIGURE 5.5 Basic rack (normal plane), coarse-pitch. (From Ref. 7.)

TABLE 5.2 Preferred Diametral Pitches
and Modules

Coarse-pitch range, P_d	Modules, m
2.0	
2.25	
2.5	2.5
3.0	
4.0	
6.0	
8.0	
10.0	
12.0	2.0
16.0	

Fine-pitch range	Modules, m
20.0	1.25
24.0	1.0
32.0	0.8
40.0	0.6
48.0	0.5
64.0	0.4
80.0	0.3
96.0	0.25
120.0	
150.0	
200.0	

Another interesting relationship can be seen in Figure 5.9. Both of the larger base circles are of the same diameter as are both of the smaller base circles whereas the distance between centers is unequal.

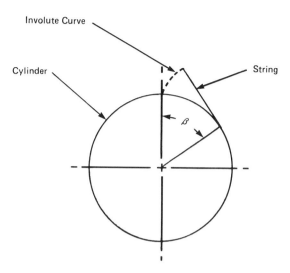

FIGURE 5.6 Inscribing an involute curve.

It can also be seen that pressure angle ϕ is different for each pair. Note, however, that pressure angle for O and for O_1 of the lefthand pair is equal for both base circles as is the pressure angle for the

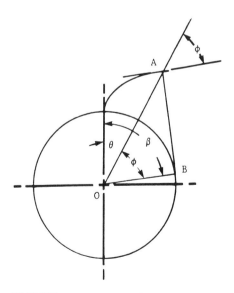

FIGURE 5.7 Geometric determination of an involute curve.

TABLE 5.3 Involute of Pressure Angle

A Pressure angle ϕ, degrees	B Tangent of ϕ	C $(\pi/180) \times \phi$	D Involute ϕ (col. B minus col. C)
14.5	0.25862	0.2530714	0.0055486
17.0	0.30573	0.2967044	0.0090256
20.0	0.36397	0.349064	0.014906
22.5	0.41421	0.392697	0.021513
25.0	0.46631	0.43633	0.02998

righthand pair. However, pressure angle increases as the distance between centers increases.

Production of Involute Profile

Actual production of an involute is illustrated by the action of a single tooth of a hob in a gear-generating machine. A hob is described as a fluted worm having straight-sided teeth. As the worm rotates in relation to the gear blank rotation, the hob is fed into the blank to a predetermined depth, thus producing involute profile gear teeth. If the proportions of the hob tooth illustrated in Figure 5.10 agreed with the values in Table 5.1, the cut gear would operate with a rack of the same gearing system (9). This can be summarized in this fashion (1) — if during the cutting the hob pitch line is tangent to the standard pitch circle of the gear:

Circular tooth thickness is one-half the hob pitch.
Circular tooth thickness is one-half the gear circular pitch.
Outside diameter of the gear will equal the standard pitch diameter +2.0 addendums.
Root diameter will be standard pitch diameter -2.0 dedendums.
Pressure angle of the gear at the standard pitch diameter will be one-half the included angle of the hob teeth.

These principles hold true for both metal and plastic gears, assuming both are cut and the plastic is held rigidly and does not move during the cutting operation. There is, however, the problem

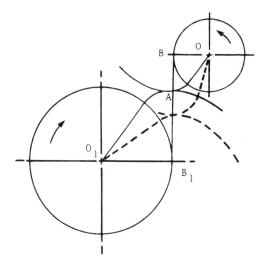

FIGURE 5.8 Involute curves in contact.

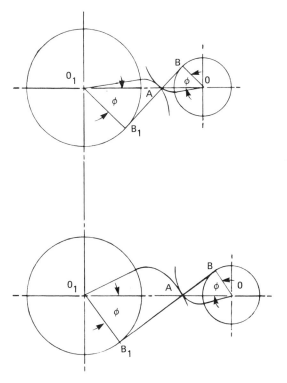

FIGURE 5.9 Mating involute curves and unequal center distances.

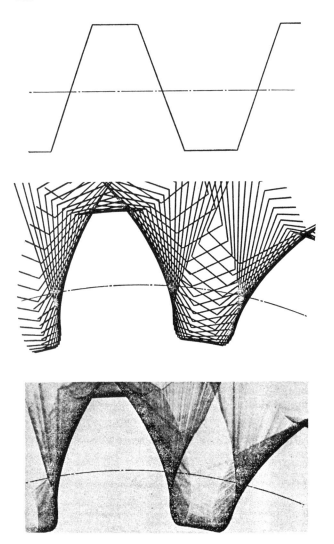

FIGURE 5.10 Gear tooth involute profile generated by straight hob tooth. (From Ref. 9.)

of tooth shrinkage if the teeth are to be molded. Molding of plastics gears is a prime advantage in production, and tooth changes due to shrinkage can be compensated for in the preparation of tooling and mold equipment. This problem has existed since the advent of the use of plastics in gear molding. L. D. Martin published his hob

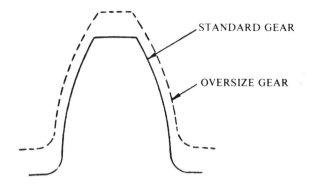

FIGURE 5.11 Improper tooth mold design. (From Ref. 1.)

pressure angle correction factor in 1954 (12,13), and several other companies had significant studies underway to eliminate the necessity for cut-and-try methods of mold production.

Molding of Involute Profile Teeth

In allowing for gear tooth shrinkage, it is not sufficient to produce a cavity as shown in Figure 5.11 and indicated by the broken line, expecting the tooth to shrink to the form as shown for the standard gear. This practice will only provide the molded gear shown by the broken line in Figure 5.12 (14). The pressure angle is incorrect, a

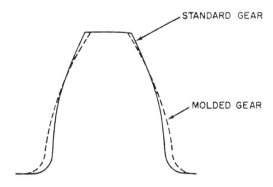

FIGURE 5.12 Standard tooth vs. molded tooth. (From Ref. 6.)

serious oversize condition exists at the gear dedendum area and root radius, and top land or tooth tip thickness is reduced.

Since gear customers expect parts to meet their drawing specifications, the molder must produce an electrode with modifications that will provide the proper dimensions after shrinkage. These considerations are perhaps the most neglected items today among the less knowledgeable gear, spline, and timing pulley plastic molders.

Mold Tool Corrections

To calculate the amount of pressure angle correction and to calculate changes to be made to the hob for cutting electrodes, the following should be used.

$$\cos \phi = \cos \frac{\phi_1}{(1+S)} \qquad (5.9)$$

$$P_c = \frac{P_d}{(1+S)} \qquad (5.10)$$

$$D_c = \frac{N}{P_c} \qquad (5.11)$$

$$C_p = \frac{\pi}{P_c} \qquad (5.12)$$

where

ϕ_1 = pressure angle of hob and gear mold
ϕ = pressure angle of molded gear
S = shrinkage rate of plastic (in./in.)
P_c = corrected diametral pitch
P_d = standard diametral pitch
N = number of teeth
C_p = corrected circular pitch
D_c = corrected pitch diameter

To mold a gear with a 14.5, 20, or 25° pressure angle, the cosine of the chosen angle should be multiplied by the value of 1.0 plus the shrinkage rate of the material selected. It can be seen that for a particular shrinkage rate and pressure angle, the hob and gear mold pressure angle correction is identical regardless of gear size. Diametral pitch, circular pitch, and pitch diameter must also be corrected but are dependent upon gear size (1,6,14).

The recommended method of producing electrodes for mold cavity production can be found in reference (1). Step-by-step details

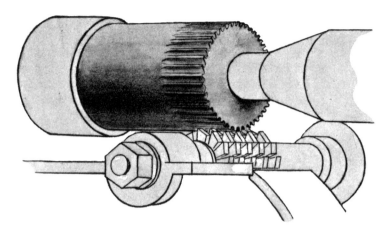

FIGURE 5.13 Generation of a gear electrode. (From Ref. 1.)

describe sources of tooth-to-tooth and composite errors, provide discussion of ways normally used to combat such errors, and illustrate
generation of a gear electrode (see Figure 5.13). It is significant
that the electrode is cut identical to the finished gear tooth form
blank in the same manner and on the same equipment as metal gears
are made. The only exception is the spark gap allowance for the
discharge machining process.

Two other methods have been used to provide corrected gear hobs
for cutting electrodes. One method is to alter a standard diametral
pitch nontopping spur gear hob by grinding to the required pressure angle. This method has often presented other problems where
diametral pitch, circular pitch, and pitch diameter are not properly
addressed and attention is only given to the pressure angle.

The second method is to produce a single-tooth hob called a fly
cutter. This special hob may be made in such a way that the cutting
teeth can be removed and replaced with other tooth cutter designs.
It may also be built to carry two teeth 180° apart. The hob is mounted
on a standard gear hobbing machine and the fly cutter rotates in relation to the electrode blank rotation the same way that a standard gear
would be cut. Figure 5.14 shows the extensive detail involved (15)
in a standard coarse-pitch hob. Similar detail must be applied to the
one- or two-tooth fly cutter. A sketch of a fly cutter is shown in
Figure 5.15.

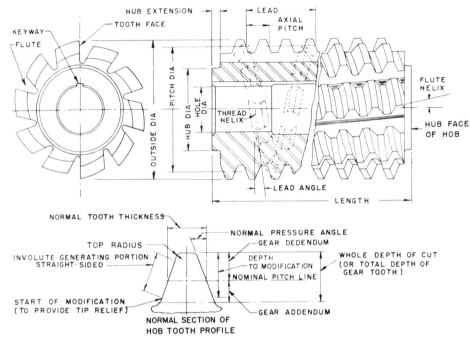

FIGURE 5.14 Hob nomenclature. (From Ref. 15.)

When producing molds for helical and other angular-type gears, close attention should be given to all angular parameters and their modifications. For helical gears, the helix angle is obtained in the gear-hobbing machine as the electrode is being cut and is therefore subject to the machine set-up process. It is recommended that experienced mold producers and gear molders be consulted whenever unusual and exceptional gearing types must be used in an application.

Tooth Undercut

It is generally agreed that tooth tip modifications and full fillet root radius should be used on all molded plastic gears. However, as the number of teeth in a gear decreases, the teeth become susceptible to severe undercut in the tooth root radius area. This problem is encountered particularly in fine-pitch metallic gearing, where the

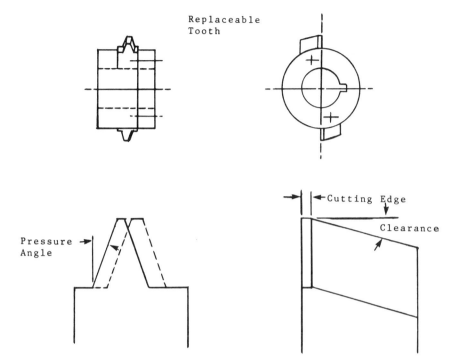

Replaceable
Tooth

Cutting Edge

Clearance

Pressure
Angle

FIGURE 5.15 Typical fly cutter design.

hob tooth tip has an increased length to allow for hob wear and for
regrinding after use. Undercutting occurs as illustrated in Figure
5.16. The Concurve® tooth was developed to minimize the severe
undercutting shown (16). Undercutting has been addressed also by
AGMA, and a modified system has been provided to deal with the problem.

The AGMA recommended tooth systems are the standard-center-
distance system and the enlarged-center-distance system. When
small numbers of teeth or special center distance situations are encountered, it is intended that long- and short-addendum gear proportions be used as specified in the enlarged-center-distance system. This also permits freedom of choice in making minor changes
in the tooth proportions to meet the special design conditions as long
as the resulting gears are fully conjugate to the basic rack. Such
changes may be indicated when there is a desire for a special contact
ratio or top land, or to reduce or eliminate undercut. It is essential

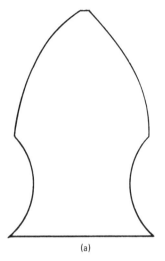
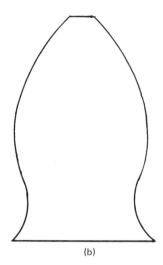

(a)   (b)

FIGURE 5.16   Involute and Concurve® tooth profile comparisons.
Standard addendum: 5 teeth, 8 diametral pitch, scale = 12X.   (a)
Involute.   (b) Concurve®.   (From Ref. 16.)

that gears with low numbers of teeth be checked for suitability, particularly in the areas of contact ratio, undercutting, and clearance.
Table 5.4 is a listing of fine-pitch tooth proportions, and Table 5.5
lists those of coarse-pitch for 20° and 25° pressure angles.   Tables
5.6 and 5.7 provide limits for numbers of teeth when using equal addendums on pinion and gear and values of addendums and tooth thicknesses for long-addendum pinions and their mating short-addendum
gears for 20° and 25° pressure angle designs, respectively.

TABLE 5.4   Tooth Proportions and Formulas for Diameter and Standard
Center Distances, in Inches, Fine-Pitch

	Tooth proportions	
Item	Spur	Helical
Addendum (a)	$1.000/P$	$1.000/P_n$
Dedendum (b)	$1.200/P + 0.002$ (min.)[a]	$1.200/P_n + 0.002$ (min.)[a]

(continued)

TABLE 5.4 (continued)

	Tooth proportions	
Item	Spur	Helical
Working depth (h_k)	2.000/P	2.000/P_n
Whole depth (h_t)	2.200/P + 0.002 (min.)[a]	2.200/P_n + 0.002 (min.)[a]
Clearance (c): Standard	0.200/P + 0.002 (min.)[a]	0.200/P_n + 0.002 (min.)[a]
Shaved or ground teeth	0.350/P + 0.002 (min.)[a]	0.350/P_n + 0.002 (min.)[a]
Tooth thickness (t)	t = 1.5708/P	t_n = 1.5708/P_n

	Formulas	
Item	Spur	Helical
Circular pitch (p)	$p = \pi D/N_G$ or $\pi d/N_p$	$P_n = n/P_n$
Pitch diameter: Pinion (d)	N_p/P	N_p/P_n cos ψ
Gear (D)	N_G/P	N_g/P_n cos ψ
Outside diameter: Pinion (d_o)	N_p + 2/P or d + 2a	$1/P_n$ (N_p/cos ψ + 2)
Gear (D_o)	N_G + 2/P or D + 2a	$1/P_n$ (N_G/cos ψ + 2)
Center distance (C)	N_G + N_p/2P or D + d/2	N_G + N_p/2P_n cos ψ

[a]The term 0.002 (min.) is added for gears with diametral pitch 20 and finer.
Key: P = transverse diametral pitch, P_n = normal diametral pitch; t_n = normal tooth thickness at pitch diameter; p_n = normal diametral pitch; ψ = helix angle; N_p = number of pinion teeth; N_G = number of gear teeth.
Source: Ref. 18.

TABLE 5.5 Tooth Proportions and Formulas for Diameters and
Center Distance, 20° and 25° Full-Depth Involute Tooth Forms —
Coarse-Pitch

Tooth proportions

$$*\text{Addendum} = \frac{1.000}{P}$$

$$\text{Dedendum} = \frac{1.250}{P} \quad (\text{Preferred})$$

$$\text{Clearance} = \frac{0.250}{P} \quad (\text{Preferred})$$

$$\text{Working depth} = \frac{2.000}{P}$$

$$\text{Whole depth} = \frac{2.250}{P} \quad (\text{Preferred})$$

$$\text{Fillet radius in basic rack} = \frac{0.300}{P}$$

$$*\text{Circular tooth thickness} - \text{basic} = \frac{\pi}{2P}$$

Formulas for diameters and center distance

$$\text{Pitch diameter} = \frac{N_G}{P}; \frac{N_P}{P}$$

$$*\text{Outside diameter} = \frac{N_G}{P} + 2a; \frac{N_P}{P} + 2a$$

$$\text{Center distance} = \frac{N_P + N_G}{2P}$$

where

N_G = number of teeth in gear
N_P = number of teeth in pinion
P = diametral pitch

*For 20° pressure angle teeth with long- and short-addendums,
see Table 5.6; for 25° pressure angle teeth, see Table 5.7.
Source: Ref. 7.

TABLE 5.6 20° Full-Depth Tooth Form — Limits for Numbers of Teeth Using Equal Addendums and Addendums and Thicknesses for Long-Addendum Pinions and Mating Short-Addendum Gears* — Coarse-Pitch

Number of teeth in pinion (N_P)	Addendum		Basic tooth thickness		Number of teeth in gear (N_G) (minimum)
	Pinion (a_P)	Gear (a_G)	Pinion (t_P)	Gear (t_G)	
10	1.468	.532	1.912	1.230	25
11	1.409	.591	1.868	1.273	24
12	1.351	.649	1.826	1.315	23
13	1.292	.708	1.783	1.358	22
14	1.234	.766	1.741	1.400	21
15	1.175	.825	1.698	1.443	20
16	1.117	.883	1.656	1.486	19
17	1.058	.942	1.613	1.529	18

*All values of addendum and thickness are for 1 diametral pitch. For other tooth sizes divide by diametral pitch. Basic tooth thickness does not include allowance for backlash. Minimum number of teeth in pinion, 18; minimum total number of teeth in pair, 36.
Source: Ref. 7.

TABLE 5.7 25° Full-Depth Tooth Form — Limits for Numbers of Teeth Using Equal Addendums and Addendums and Tooth Thicknesses for Long-Addendum Pinions and Mating Short-Addendum Gears* — Coarse-Pitch

Number of teeth in pinion (N_P)	Addendum		Basic tooth thickness		Number of teeth in gear (N_G) (minimum)
	Pinion (a_P)	Gear (a_G)	Pinion (t_P)	Gear (t_G)	
10	1.184	.816	1.742	1.399	15
11	1.095	.905	1.659	1.482	14

*All values of addendums and thickness are for 1 diametral pitch. For other tooth sizes divide by diametral pitch. Basic tooth thickness does not include allowance for backlash. Minimum number of teeth in pinion, 12; minimum total number of teeth in pair, 24.
Source: Ref. 7.

118   *Tooth Elements*

TABLE 5.8 Basic Tooth Dimensions (Fine-Pitch)

Diametral pitch (spur, P; helical, P_n)	Circular pitch (spur, p; helical, p_n)	Circular tooth (thickness, t; helical, t_n)	Addendum (a)	Clearance (c)	Whole depth (h_t)
20[a]	0.1571	0.0785	0.0500	0.0120	0.1120
22	0.1428	0.0714	0.0455	0.0111	0.1021
24[a]	0.1309	0.0654	0.0417	0.0103	0.0937
26	0.1208	0.0604	0.0385	0.0097	0.0867
28	0.1122	0.0561	0.0357	0.0091	0.0805
30	0.1047	0.0524	0.0333	0.0087	0.0753
32[a]	0.0982	0.0491	0.0313	0.0083	0.0708
34	0.0924	0.0462	0.0294	0.0079	0.0667
36	0.0873	0.0436	0.0278	0.0076	0.0631
38	0.0827	0.0413	0.0263	0.0073	0.0599
40[a]	0.0785	0.0393	0.0250	0.0070	0.0570
44	0.0714	0.0357	0.0227	0.0065	0.0520
48[a]	0.06545	0.03272	0.0208	0.0062	0.0478
52	0.06042	0.03021	0.0192	0.0058	0.0442
56	0.05610	0.02805	0.0179	0.0056	0.0413
60	0.05236	0.02618	0.0167	0.0053	0.0387
64[a]	0.04909	0.02454	0.0156	0.0051	0.0364
72	0.04363	0.02182	0.0139	0.0048	0.0326
80[a]	0.03927	0.01964	0.0125	0.0045	0.0295
96[a]	0.03272	0.01636	0.0104	0.0041	0.0249
120[a]	0.02618	0.01309	0.0083	0.0037	0.0203

[a]Preferred pitches.

Basic tooth dimensions for fine-pitch, coarse-pitch, and metric module gears are tabulated in Tables 5.8, 5.9, and 5.10. The gearing industry generally recognizes and recommends 20° pressure angle as standard for most applications.

TABLE 5.9 Basic Tooth Dimensions (Coarse-Pitch)

Diametral pitch (spur, P; helical, P_n)	Circular pitch (spur, p; helical, p_n)	Circular tooth (thickness, t; helical, t_n)	Addendum (a)	Clearance[a] (c)	Whole[a] depth (h_t)
1	3.1416	1.5708	1.0000	0.2500	2.2500
2	1.5708	0.7854	0.5000	0.1250	1.1250
3	1.0472	0.5236	0.3333	0.0833	0.7500
4	0.7854	0.3927	0.2500	0.0625	0.5625
5	0.6283	0.3142	0.2000	0.0500	0.4500
6	0.5236	0.2618	0.1666	0.0416	0.3750
7	0.4488	0.2244	0.1429	0.0357	0.3214
8	0.3927	0.1963	0.1250	0.0312	0.2812
9	0.3491	0.1745	0.1111	0.0278	0.2500
10[a]	0.3142	0.1571	0.1000	0.0250	0.2250
12[a]	0.2618	0.1309	0.0833	0.0208	0.1875
14	0.2244	0.1122	0.0714	0.0179	0.1607
16[a]	0.1963	0.0982	0.0625	0.0156	0.1406
18	0.1745	0.0873	0.0555	0.0139	0.1250

[a]Preferred Values.

The following guidelines should be considered:

1. Where greater strength and wear resistance are required, a 25° pressure angle may be desirable. Close control of center distance tolerance is absolutely necessary in applications where backlash is critical if the 25° pressure angle is used.
2. When angular position or backlash are critical and where both pinion and gear have relatively large numbers of teeth, a 14 1/2° pressure angle may be desirable. However, in general, pressure angles of less than 20° require a greater amount of modification to avoid undercut problems and are limited to larger total numbers of teeth in gear and pinion when operating on a standard center distance.

TABLE 5.10 Basic Tooth Dimensions for Metric Modules

Module (spur, m; helical, m_n)	Circular pitch (spur, p; helical, p_n)	Circular tooth (thickness, t; helical, t_n)	Addendum (a)	Whole depth (h_t)	Dedendum (b)	Clearance (c)
5	15.708	7.854	5.0	11.25	6.25	1.25
4	12.5664	6.2832	4.0	9.0	5.0	1.0
3.2	10.2102	5.1051	3.25	7.3125	4.0625	0.8
2.5[a]	7.854	3.927	2.5	5.625	3.125	0.625
2.0[a]	6.2832	3.1416	2.0	4.5	2.5	0.5
1.5	4.7124	2.3562	1.5	3.375	1.875	0.375
1.25[a]	3.927	1.9635	1.25	2.8125	1.5625	0.3125
1.0[a]	3.1416	1.5708	1.0	2.25	1.25	0.25
0.9	2.8274	1.4137	0.9	2.025	1.125	0.225
0.8[a]	2.5133	1.25665	0.8	1.8	1.0	0.2
0.75	2.3562	1.1781	0.75	1.6875	0.9375	0.1875
0.7	2.1991	1.0996	0.7	1.575	0.875	0.175
0.6[a]	1.885	0.9425	0.6	1.35	0.75	0.15
0.5[a]	1.5708	0.7854	0.5	1.125	0.625	0.125
0.4[a]	1.2566	0.6283	0.4	0.9	0.5	0.1
0.3[a]	0.9425	0.4712	0.3	0.675	0.375	0.075
0.25[a]	0.7854	0.3927	0.25	0.5625	0.3125	0.0625

[a]Preferred modules.
All dimensions in millimeters, mm. Data above the line correspond
to diametral pitch coarse-pitch series. Data below the line correspond to diametral pitch fine-pitch series. Data in the metric module
system are not directly interchangeable with data in the diametral
pitch system.
Source: ISO Standard 53, ISO Recommendation R54.

TABLE 5.11 Minimum Number of Pinion Teeth vs. Pressure Angle Having No Objectionable Undercut

Helix angle, ψ (degrees)	Minimum number of teeth to avoid undercut (normal pressure angle, ϕn, degrees)		
	14 1/2	20	25
0	32	24	15
5	32	24	15
10	31	23	14
15	29	22	14
20	27	20	13
23	26	19	12
25	25	18	11
30	22	16	10
35	19	13	9
40	15	11	7
45	12	9	6

3. Equal-addendum teeth are used for general purpose applications where the number of teeth are equal to or exceed the minimum number shown in Table 5.11 in relation to pressure angle and helix angle.
4. To insure smooth operation, contact ratio should never be less than 1.2; therefore long- and short-addendum proportions should be used.

It should be noted that the tables presented here do not include an allowance for backlash when the gears are meshed at standard center distance. In equal addendum gearing, the teeth of both members are generally reduced in thickness by an equal amount to provide backlash. Where pinions have small numbers of teeth, all of the tooth thickness reduction for backlash should be applied to the gear member. Tables 5.12 and 5.13 are provided for guidance.

TABLE 5.12 Center Distance and Long- and Short-Addendum Relations

Pinion	Gear	Center distance
Enlarged	Diameter and tooth thickness decreased by amount of pinion enlargement	Standard
Enlarged	Enlarged pinion	Greater than standard
Enlarged	Standard	Greater than standard

Standard Center-Distance System (Long and Short Addendums, Operating Pressure Angle Constant)

Pinion	Gear	Center distance
Enlarged	Outside diameter, root diameter decreased the amount pinion increased	Standard

Enlarged Center-Distance System

Pinion	Gear	Center distance
Enlarged	Enlarged pinion	Greater than standard[a]
Enlarged	Standard	Greater than standard[a]

[a]See discussion on backlash treatment in Chapter 8.

Generally, spur pinion enlargement will follow this procedure for 20° pressure angle and diametral pitches 20 through 120:

1. Tooth thickness and outside diameter are obtained by dividing column 2 and column 3 values of Table 5.14 by the diametral pitch.
2. Pinion dedendum is determined by dividing column 5 of Table 5.14 by diametral pitch and adding the increment "Δ" of column 9 from Table 5.15.
3. Enlarged center distance is obtained by formula under Figure 5.17 using center distance factor K from column 9 of Table 5.14.

TABLE 5.13 Advantages and Disadvantages of Center-Distance Systems

---

Standard-Center-Distance System

Advantages: No change in center distance
                Operating pressure angle standard
                Slight increase in contact ratio when center distance
                increased

Disadvantages: Gear and pinion both nonstandard
                   2 pinions, 20° pressure angle; fewer than 24 teeth,
                   requiring enlargement, do not mesh satisfactorily
                   Usually cannot be used in trains with idler gears

---

Enlarged-Center-Distance System

Advantages: Only pinions need be changed from standard
                Pinions with less than 24 teeth will operate satisfactorily

Disadvantages: Center distances must be enlarged
                   Slight increase in operating pressure angle
                   Slightly smaller contact ratio

---

4.   Short addendum gears should be designed in accordance with
     Table 5.16. Dedendum is determined by dividing the value in
     column 4 by diametral pitch and adding the increment "Δ" shown
     in column 9 of Table 5.15.

Note that Figure 5.19 provides the center distance factor for 25° pressure angle enlarged pinions.

Figure 5.18 provides recommended enlargement and calculation procedures for helical pinions of 20° normal pressure angle. Enlargement
to avoid undercut is provided in graph form because of the wide range
of helix angles in common use. When using enlarged helical pinions,
the mating gear must be reduced in diameter or the center distance
increased to correspond with the enlargement, as in spur gears.

The rationale behind the enlargement values is based on both operational and manufacturing requirements. Manufacturing difficulties
arise from nonstandard hobs, attempts to start the involute tooth form
close to the base circle, small radii on cutting tools that wear quickly,

TABLE 5.14 Tooth Proportions Recommended for Enlarging 20 Degree Pressure Angle Pinions of Small Numbers of Teeth

	Pinion dimensions							Enlarged C.D. system pinion mating with standard gear		
1	2	3	4	5	6	7	8	9	10	11
Number of teeth (n)	Outside diameter	Addendum	Basic tooth thickness	Dedendum based on 20 pitch	Form diameter	Roll angle to form diam.	Top land	Center distance factor K^b	Contact ratio two equal pinions	Contact ratio with 24 tooth gear
7	10.0102	1.5051	2.14114	0.4565	6.6307	7.28°	0.2750	0.7835	0.697	1.003
8	11.0250	1.5125	2.09854	0.5150	7.5638	6.37°	0.2750	0.7250	0.792	1.075
9[a]	12.0305	1.5152	2.05594	0.5735	8.4984	5.66°	0.2750	0.6665	0.893	1.152
10	13.0279	1.5140	2.01355	0.6321	9.4340	5.10°	0.2750	0.6080	0.982	1.211
11	14.0304	1.5152	1.97937	0.6787	10.3761	5°	0.2750	0.5613	1.068	1.268
12	15.0296	1.5148	1.94703	0.7232	11.3194	5°	0.2750	0.5168	1.151	1.322
13	15.9448	1.4724	1.91469	0.7676	12.2627	5°	0.3400	0.4724	1.193[b]	1.353

n	Do_p	a_p	t_p	b_p	Df_p	Ef_p	to_p	K	m_p	m_p
14	16.8560	1.4280	1.88235	0.8120	13.2060	5°	0.3995	0.4280	1.232	1.381
15	17.7671	1.3836	1.85001	0.8564	14.1492	5°	0.4512	0.3836	1.270	1.408
16	18.6782	1.3391	1.81766	0.9009	15.0925	5°	0.4968	0.3391	1.323	1.434
17	19.5894	1.2947	1.78532	0.9453	16.0358	5°	0.5370	0.2947	1.347	1.458
18	20.5006	1.2503	1.75298	0.9897	16.9791	5°	0.5727	0.2503	1.385	1.482
19	21.4116	1.2058	1.72064	1.0342	17.9224	5°	0.6046	0.2058	1.423	1.505
20	22.3228	1.1614	1.68839	1.0786	18.8657	5°	0.6330	0.1614	1.461	1.527
21	23.2340	1.1170	1.65595	1.1230	19.8089	5°	0.6585	0.1170	1.498	1.548
22	24.1450	1.0725	1.62361	1.1675	20.7522	5°	0.6814	0.0725	1.536	1.568
23	25.0561	1.0281	1.59127	1.2119	21.6954	5°	0.7020	0.0281	1.574	1.588
24	26.0000	1.0000	1.57080	1.2400	22.6388	5°	0.7155	0.0000	1.602	1.602

[a]Caution is to be exercised in the use of gears with numbers of teeth appearing in the shaded area. They should be checked for suitability, particularly in the areas of contact ratio, center distance, clearance and tooth strength.
[b]See Figure 5.18.
Source: Ref. 18.

TABLE 5.15 Tooth Dimensions for 20° Pressure Angle Fine-Pitch Gears[a]

1	2	3	4	5	6	7	8	9
Diametral pitch	Circular pitch	Standard circular thickness	Working depth	Whole depth	Clearance	Standard addendum	Standard dedendum	Δ
20	0.15708	0.07854	0.1000	0.1120	0.0120	0.0500	0.0620	.0000
24	0.13090	0.06545	0.0833	0.0937	0.0104	0.0417	0.0520	.0004
32	0.09818	0.04909	0.0625	0.0708	0.0083	0.0313	0.0395	.0007
40	0.07854	0.03972	0.0500	0.0570	0.0070	0.0250	0.0320	.0010
48	0.06545	0.03272	0.0417	0.0478	0.0062	0.0208	0.0270	.0012
64	0.04909	0.02454	0.0312	0.0364	0.0051	0.0156	0.0208	.0015
72	0.04363	0.02182	0.0278	0.0326	0.0048	0.0139	0.0187	.0015
80	0.03927	0.01964	0.0250	0.0295	0.0045	0.0125	0.0170	.0015
96	0.03272	0.01636	0.0208	0.0249	0.0041	0.0104	0.0145	.0016
120	0.02618	0.01309	0.0167	0.0203	0.0037	0.0083	0.0120	.0017

[a]Helical teeth are conjugated in the normal plan to the basic rack. Tooth proportions for both the 14 1/2-degree pressure angle system and the 25-degree pressure angle system are the same as those used by the 20-degree pressure angle system shown in Table 5.4.
All dimensions are given in inches.
Δ = increase in dedendum over the values determined by dividing the dedendums shown in Tables 5.14, 5.16, 5.18, and 5.19 by diametral pitch. The corrected dedendums would be conjugate with the basic rack dedendums of $1.2/P_d$ + .002, where P_d = diametral pitch, p = circular pitch, t = standard circular thickness, h_K = working depth, h_t = whole depth, c = clearance, a = standard addendum, and b = standard dedendum.
Source: Ref. 18.

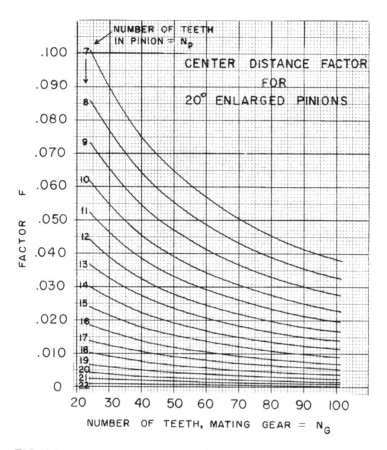

FIGURE 5.17   Center distance factor F for 20° pressure angle enlarged pinions.   Center distance $C = 1/P_d$ $(N_p + N_G/2 + K - F)$.
Note: The center distance as calculated above is based on the general formulas: $C = (N_p + N_G/2P_d)$ $(\cos \phi_1/\cos \phi_2)$ where inv $\phi_2 =$
inv $\phi_1 + (T_P + T_G - p)P_d/N_p + N_G$, P_d = diametral pitch, K = center
distance factor (Table 5.14), ϕ_1 = 20-degree pressure angle, T_p =
tooth thickness of pinion, T_G = tooth thickness of gear, p = circular
pitch. (From Ref. 18.)

and difficulties in holding tolerances. Application concerns are proper
meshing of the pinion, adequate tooth top land, and lack of undercut.
    Breur (17) developed the enlargement data that was later ap-
proved by the AGMA (18). The basis for the enlarged pinion proportions are:

TABLE 5.16   20° Pressure Angle Enlarged Pinions

			Standard center distance system (long and short addendum)		
			Gear dimensions		
1	2	3	4	5	6
Number of teeth in pinion	Addendum	Basic tooth thickness	Dedendum based on 20 pitch	Recommended minimum number of teeth (N)	Contact ratio, n, mating with N
7	0.2165	1.00045	2.0235	42	1.079[a]
8	0.2750	1.04305	1.9650	40	1.162[a]
9	0.3335	1.08565	1.9065	39	1.251
10	0.3921	1.12824	1.8479	38	1.312
11	0.4387	1.16222	1.8013	37	1.371
12	0.4832	1.19456	1.7568	36	1.427
13	0.5276	1.22690	1.7124	35	1.457
14	0.5720	1.25924	1.6680	34	1.483
15	0.6164	1.29158	1.6236	33	1.507
16	0.6609	1.32393	1.5791	32	1.528
17	0.7053	1.35627	1.5347	31	1.546
18	0.7497	1.38861	1.4903	30	1.561
19	0.7942	1.42095	1.4458	29	1.574
20	0.8386	1.45320	1.4014	28	1.584
21	0.8830	1.48564	1.3570	27	1.592
22	0.9275	1.51798	1.3125	26	1.598
23	0.9719	1.55032	1.2681	25	1.601
24	1.0000	1.57080	1.2400	24	1.602
n	aG	tG	bG	N	mp

[a]Special attention must be given to avoid contact ratios below 1.2.
Source: Ref. 18.

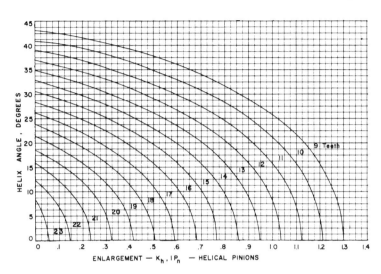

ENLARGEMENT — K_h, $1P_n$ — HELICAL PINIONS

Example

Given: 12 teeth, 32 normal diametral pitch, 20-degree normal pressure angle and 18-degree helix angle.

$P = P_n \times \cos \psi$

$P = 32 \times 0.95106 = 30.4339$

$d = \dfrac{12}{30.4339} = 0.3943$

$K_h = .851$ (from graph)

$d_o = 0.3943 + \dfrac{2 + .851}{32} = .4834$

Graph Derivation

$$\tan \phi_t = \frac{\tan \phi_n}{\cos \psi}$$

$$2\Delta C = K_h = \text{enlargement} = 2.1 - \frac{n}{\cos \psi} \left[\sin \phi_t - \cos \phi_t \tan 5° \right] \sin \phi_t$$

ϕ_n = normal pressure angle

ϕ_t = transverse pressure angle

FIGURE 5.18 Recommended enlargement for helical pinions. Note: These data are based on the use of hobs having sharp corners at their top lands. Pinions cut by sharper cutters may not require as much modification. (From Ref. 18.)

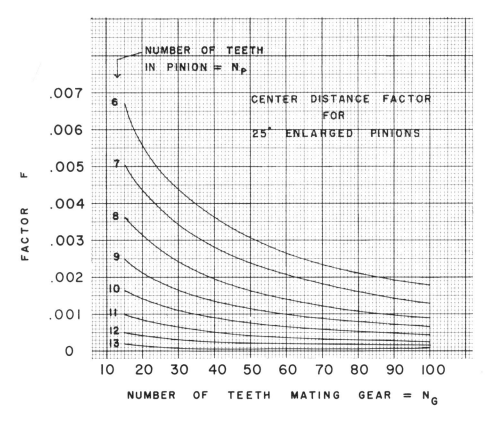

FIGURE 5.19 Center distance factor F for 25° pressure angle enlarged pinions. See Figure 5.17 for center distance calculation. ϕ_1 = 25-degree pressure angle. (From Ref. 18.)

1. Pinions with 12 to 23 teeth are enlarged so that a standard tooth thickness rack with addendum 1.05/pitch will start contact 5° of roll above the base radius. This allows for center distance variation and eccentricity of the mating gear outside diameter. The 5° roll angle avoids the fabrication of the involute in the area near the base circle.

2. Pinions with fewer than 12 teeth are enlarged to the extent that the highest point of undercut coincides with the start of contact with the standard rack described above. The height of undercut considered was that produced by a sharp-cornered 120-pitch hob.

3. Pinions with fewer than 12 teeth are truncated to provide a top land of 0.275/pitch.

4. The form diameters and the roll angles to form diameter give the values that should be met with a standard hob when generating the tooth thicknesses shown in the tables. These form diameters provide more than enough length of involute profile for any mating gear smaller than a rack. However, since these form diameters are based on the generation with standard hobs, they should impose little or no hardship on manufacturers except in the cases of the most critical quality levels. In these cases, form diameter specification and master gear design should be based on actual mating conditions. Tables 5.17 and 5.18 data cover 14.5° and 25° pressure angle full-depth involute tooth proportions that differ from those tabulated previously. Tooth proportions for 14.5°, 20°, and 25° pressure-angle systems are given in Tables 5.1 and 5.4. Table 5.17 provides data for 14.5° pressure-angle standard center-distance and enlarged center-distance systems. Table 5.18 and Figure 5.19 cover the 25° pressure angle enlarged, and Table 5.19 the standard-center-distance system.

PGT Tooth System

Another system of merit has been developed by Plastics Gearing Technology, Inc. (1) that addresses the tooth undercut problem, effects of tooth deflection using tooth tip relief, and reduction of fatigue stress using full fillet root radii. The system is referred to as the "PGT system" and the tooth forms as the "PGT tooth forms."

Figure 5.20 is the basic tooth form of the PGT system. Gears having this tooth form will function with a mating gear having either the fine- or coarse-pitch standard tooth form. The tooth forms shown in Figures 5.21-5.23 are modifications of the basic PGT tooth form with the advantages described but with longer teeth than standard, as can be seen by the specification of ht. Designated as PGT-2 through PGT-4, they are generally recommended for fine-pitch instrument gears such as electric clocks, control mechanisms, meters, cameras, and similar items.

Calculations show that when proper compensations are made for temperature rise and moisture absorption, the opposite conditions can cause gears with standard design to separate.

Table 5.20 is used to determine the outside diameter using the PGT system. After calculation of the outside diameter, the following formula should be used, and the lesser of the two outside diameter values used. This insures sufficient tooth top land.

TABLE 5.17 14.5° Pressure Angle Involute Fine-Pitch System for Modified 31-Teeth or Less Spur Pinions

	Pinion dimensions		Standard center distance system (long and short addendum) Gear dimensions[a]				Enlarged center-distance system standard mating gear diameter[b]	
1	2	3	4	5	6	7	8	9
Number of teeth (n)	Outside diameter (D_o)	Circular tooth thickness at standard pitch diameter ($\Delta t_p = \Delta d \tan \phi$)	Decrease in standard outside diameter	Circular tooth thickness at standard pitch diameter ($\Delta t_G = \Delta D \tan \phi$)	Recommended minimum number of teeth (N)	Contact ratio, n, mating with N_3	Increase over standard center distance	Contact ratio two equal pinions[c]
10	13.3731	1.9259	1.3731	1.2157	54	1.831	.6866	1.053
11	14.3104	1.9097	1.3104	1.2319	53	1.847	.6552	1.088
12	15.2477	1.8935	1.2477	1.2481	52	1.860	.6239	1.121
13	16.1850	1.8773	1.1850	1.2643	51	1.873	.5925	1.154[d]
14	17.1223	1.8611	1.1223	1.2805	50	1.885	.5612	1.186
15	18.0597	1.8448	1.0597	1.2967	49	1.896	.5299	1.217
16	18.9970	1.8286	.9970	1.3130	48	1.906	.4985	1.248
17	19.9343	1.8124	.9343	1.3292	47	1.914	.4672	1.278

18	20.8716	1.7962	.8716	1.3454	46	1.922	.4358	1.307
19	21.8089	1.7800	.8089	1.3616	45	1.929	.4045	1.336
20	22.7462	1.7638	.7438	1.3778	44	1.936	.3731	1.364
21	23.6835	1.7476	.6835	1.3940	43	1.942	.3418	1.392
22	24.6208	1.7314	.6208	1.4102	42	1.948	.3104	1.419
23	25.5581	1.7151	.5581	1.4265	41	1.952	.2791	1.446
24	26.4954	1.6989	.4954	1.4427	40	1.956	.2477	1.472
25	27.4328	1.6827	.4328	1.4589	39	1.960	.2164	1.498
26	28.3701	1.6665	.3701	1.4751	38	1.963	.1851	1.524
27	29.3074	1.6503	.3074	1.4913	37	1.965	.1537	1.549
28	30.2447	1.6341	.2448	1.5075	36	1.967	.1224	1.573
29	31.1820	1.6179	.1820	1.5237	35	1.969	.0910	1.598
30	32.1193	1.6017	.1193	1.5399	31	1.970	.0597	1.622
31	33.0566	1.5854	.0566	1.5562	33	1.971	.0283	1.646

a To maintain standard center distances when using enlarged pinions, the mating gear dimensions must be decreased by the amount of the pinion enlargement.

b If mating gears are made with standard proportions, the center distance must be increased as shown.

c Nominal values; will vary due to the effects of tolerances.

All dimensions are given in inches. Tabular values are in inches for 1 diametral pitch. For other pitches, divide tabular values by the diametral pitch.

d Special attention must be given to avoid contact ratios below 1.2.

Source: Ref. 18.

TABLE 5.18 Tooth Proportions Recommended for Enlarging 25° Pressure Angle Pinions of Small Numbers of Teeth

	Pinion dimensions							Enlarged C.D. system pinion mating with standard gear		
1	2	3	4	5	6	7	8	9	10	11
Number of teeth n	Outside diameter D_o	Addendum a	Basic tooth thickness t	Dedendum based on 20 pitch b	Form diameter D_f	Roll angle to form diam.	Top land	Center distance factor K^b	Contact ratio two equal pinions	Contact ratio with 15 tooth gear
6	8.7645	1.3822	2.18362	0.5829	5.4588	7.12°	0.275	0.6571	0.696	0.954
7	9.7253	1.3626	2.10029	0.6722	6.3801	6.10°	0.275	0.5678	0.800	1.026[a]
8	10.6735	1.3368	2.01701	0.7616	7.2819	5.34°	0.275	0.4784	0.904	1.094[a]
9	11.6203	1.3102	1.9411	0.8427	8.1879	5°	0.275	0.3971	1.003[a]	1.156
10	12.5691	1.2846	1.87345	0.9155	9.0977	5°	0.275	0.3245	1.095	1.211
11	13.5039	1.2520	1.80579	0.9880	10.0075	5°	0.2807	0.2520	1.183	1.261
12	14.3588	1.1794	1.73813	1.0606	10.9172	5°	0.3478	0.1794	1.231	1.290
13	15.2138	1.1069	1.67047	1.1331	11.8270	5°	0.4034	0.1069	1.279	1.317
14	16.0686	1.0343	1.60281	1.2057	12.7368	5°	0.4500	0.0343	1.328	1.343
15	17.0000	1.0000	1.57030	1.2400	13.6465	5°	0.4743	0.0000	1.358	1.358

[a] Special attention must be given to avoid contact ratios below 1.2.
[b] See Figure 5.19.
Source: Ref. 18.

TABLE 5.19 25° Pressure Angle Enlarged Pinions

	Standard center distance system (long and short addendum) Gear dimensions				
1	2	3	4	5	6
Number of teeth in pinion (n)	Addendum (a_G)	Basic tooth thickness (t_G)	Dedendum based on 20 pitch (b_G)	Recom- mended minimum number of teeth (N)	Contact ratio, n, mating with N (m_p)
6	0.3429	0.95797	1.8971	24	1.030
7	0.4322	1.04130	1.8078	23	1.108^a
8	0.5216	1.12459	1.7184	22	1.177
9	0.6029	1.20048	1.6371	20	1.234
10	0.6755	1.26814	1.5645	19	1.282
11	0.7480	1.33581	1.4920	18	1.322
12	0.8206	1.40346	1.4194	17	1.337
13	0.8931	1.47112	1.3469	16	1.347
14	0.9657	1.53878	1.2743	15	1.352
15	1.0000	1.57080	1.2400	15	1.358

[a]Special attention must be given to avoid contact ratios below 1.2.
Source: Ref. 18.

TABLE 5.20 PGT System — Outside Diameter

Tooth form	Outside diameter, D_O
PGT-1	$D_O = \frac{1}{P} (N - 2.3158) + (2.7475 \times t)$
PGT-2	$D_O = \frac{1}{P} (N - 2.0158) + (2.7475 \times t)$
PGT-3	$D_O = \frac{1}{P} (N - 1.8158) + (2.7475 \times t)$
PGT-4	$D_O = \frac{1}{P} (N - 1.6158) + (2.7475 \times t)$

P = diametral pitch; N = number of teeth; t = tooth thickness.
Source: Ref. 1.

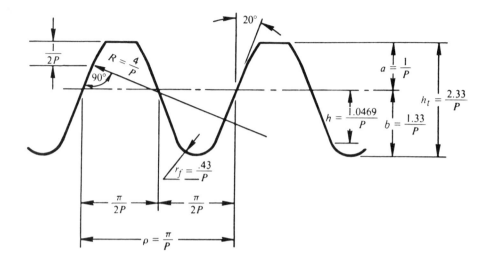

P = diametral pitch ρ = circular pitch   20° = pressure angle
a = addendum b = dedendum h_t = whole depth r_f = root radius
h = depth of straight portion of dedendum to point of tangency with root radius.

FIGURE 5.20 PGT system basic tooth form – PGT-1. (From Ref. 1.)

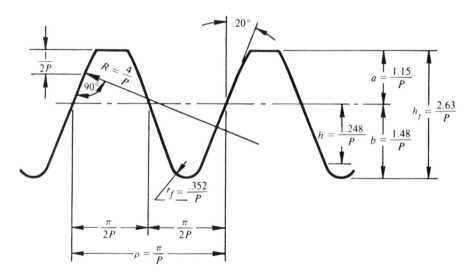

FIGURE 5.21 PGT-2 tooth form (ht = 2.63/P). (From Ref. 1.)

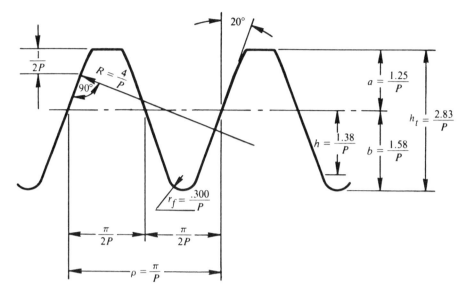

FIGURE 5.22 PGT-3 tooth form (ht = 2.83/P). (From Ref. 1.)

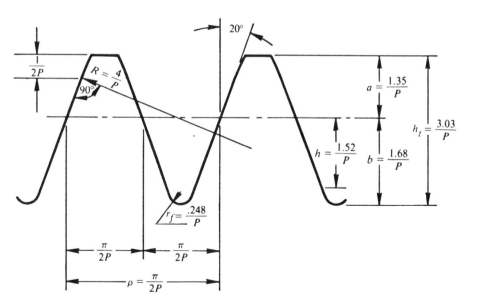

FIGURE 5.23 PGT-4 tooth form (ht = 3.03/P). (From Ref. 1.)

Tooth Elements

TABLE 5.21 PGT System — Root Diameter

Tooth form	Root diameter, D_R
PGT-1	$D_R = \frac{1}{P}(N - 6.9758) + (2.7475 \times t)$
PGT-2	$D_R = \frac{1}{P}(N - 7.2758) + (2.7475 \times t)$
PGT-3	$D_R = \frac{1}{P}(N - 7.4758) + (2.7475 \times t)$
PGT-4	$D_R = \frac{1}{P}(N - 7.6758) + (2.7475 \times t)$

P = diametral pitch; N = number of teeth; t = tooth thickness.
Source: Ref. 1.

$$D_0 (max) = \frac{0.9396926\, N}{1.017P \cos \phi_1} \qquad (5.13)$$

where

D_0 = outside diameter
P = diametral pitch
N = number of teeth
t = circular tooth thickness on standard pitch circle
$D_0 (max)$ = maximum allowable outside diameter
ϕ_1 = angle whose involute is

$$\frac{tP}{N} + 0.01490438 \qquad (5.14)$$

Table 5.21 can be used for determination of the tooth root diameter based on the tooth form selected. (D_R = root diameter.) PGT illustrates the need for tooth design modifications using Figure 5.24. The two gears have the same: number of teeth, diametral pitch, and PGT-1 tooth form. Tooth (a) has standard tooth thickness of 3.1415926/2P, whereas tooth (b) has an increased tooth thickness, involute curvature, and reduced undercut. To insure adequate involute profiles and avoid undercutting, PGT tooth forms should be designed to the minimum circular tooth thickness values given in Table 5.22. This is especially important for gears with small numbers of teeth.

TABLE 5.22 PGT System — Minimum Circular
Tooth Thickness

Tooth form	Minimum circular tooth thickness, t
PGT-1	$t = \dfrac{2.3329 - (0.0426 \times N)}{P}$
PGT-2	$t = \dfrac{2.4793 - (0.0426 \times N)}{P}$
PGT-3	$t = \dfrac{2.5768 - (0.0426 \times N)}{P}$
PGT-4	$t = \dfrac{2.6751 - (0.0426 \times N)}{P}$

t = minimum circular tooth thickness; N = number of teeth; P = diametral pitch.
Source: Ref. 1.

Tables 5.23 to 5.26 give values for each PGT tooth form for gears of 6 teeth up to a number where no undesirable undercutting is present at the standard tooth thickness. If a gear has a tooth thickness less than standard and a number of teeth more than the maximum

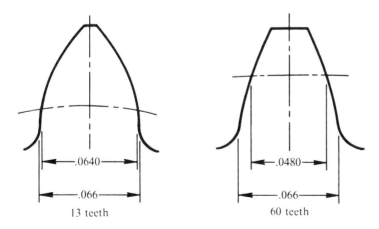

FIGURE 5.24 PGT system — equal strength teeth. (From Ref. 1.)

TABLE 5.23 PGT System – Tooth Form for Small Numbers of Teeth
(PGT-1)[a]

Number of teeth, n	Minimum circular tooth thickness, t	Outside diameter, D_O	Root diameter, D_R
6	2.0773	8.9254	4.7316
7	2.0347	9.9477	5.6145
8	1.9921	10.9578	6.4975
9	1.9495	11.9577	7.3805
10	1.9069	12.9234	8.2634
11	1.8643	13.8064	9.1464
12	1.8217	14.6893	10.0293
13	1.7791	15.5723	10.9123
14	1.7365	16.4553	11.7952
15	1.6939	17.3382	12.6782
16	1.6513	18.2212	13.5611
17	1.6087	19.1041	14.4441
18	1.5708	20.0000	15.3400

[a]Values are for 1.0 diametral pitch. For other values, divide by
desired diametral pitch.
Source: Ref. 1.

given in the appropriate table, the equation from Table 5.22 must be
used to insure that the tooth thickness is not less than the allowable
minimum.

In plastics power gearing, it is important to design both pinion and
gear with balanced tooth strength. For gears having the PGT-1 tooth
form, the following formulas are to be used.

When the pinion and gear both have fewer than 35 teeth:

$$t_1 = \frac{2.3329 - (0.0219 \times N_1)}{P} \qquad (a) \qquad\qquad (5.15)$$

$$t_2 = \frac{2.3329 - (0.0219 \times N_2)}{P} \qquad (b) \qquad\qquad (5.16)$$

TABLE 5.24 PGT System — Tooth Form for Small Numbers of Teeth
(PGT-2)[a]

Number of teeth, n	Minimum circular tooth thickness, t	Outside diameter, D_O	Root diameter, D_R
6	2.2237	9.0970	4.8338
7	2.1811	10.1247	5.7168
8	2.1385	11.1398	6.5997
9	2.0959	12.1446	7.4827
10	2.0533	13.1405	8.3656
11	2.0107	14.1287	9.2486
12	1.9681	15.1100	10.1316
13	1.9255	16.0852	11.0145
14	1.8829	17.0549	11.8975
15	1.8403	18.0196	12.7804
16	1.7977	18.9234	13.6634
17	1.7551	19.8064	14.5463
18	1.7125	20.6893	15.4293
19	1.6699	21.5723	16.3123
20	1.6273	22.4552	17.1952
21	1.5847	23.3382	18.0782
22	1.5708	24.3000	19.0400

[a]Values are for 1.0 diametral pitch. For other values, divide by desired diametral pitch.
Source: Ref. 1.

When the pinion has fewer than 35 teeth and the gear has 35 teeth or more:

$$t_1 = \frac{2.3329 - (0.0219 \times N_1)}{P} \qquad\text{(a)} \qquad (5.17)$$

$$t_2 = \frac{N_2}{P}\left(\frac{2.1922 - (0.0066 \times N_1)}{N_2 - 2.0938} + \text{inv } \phi_2 - 0.01490438\right) \qquad\text{(b)} \qquad (5.18)$$

TABLE 5.25 PGT System — Tooth Form for Small Numbers of Teeth
(PGT-3)[a]

Number of teeth, n	Minimum circular tooth thickness, t	Outside diameter, D_O	Root diameter, D_R
6	2.3212	9.2103	4.9017
7	2.2786	10.2413	6.1447
8	2.2360	11.2597	6.6676
9	2.1934	12.2677	7.5466
10	2.1508	13.2666	8.4335
11	2.1082	14.2577	9.3165
12	2.0656	15.2419	10.1994
13	2.0230	16.2200	11.0824
14	1.9804	17.1924	11.9653
15	1.9378	18.1598	12.8483
16	1.8952	19.1226	13.7313
17	1.8526	20.0810	14.6142
18	1.8100	21.0355	15.4792
19	1.7674	21.9863	16.3801
20	1.7248	22.9231	17.2631
21	1.6822	23.8061	18.1460
22	1.6396	24.6890	19.0290
23	1.5970	25.5720	19.9120
24	1.5708	26.5000	20.8400

[a]Values are for 1.0 diametral pitch. For other values, divide by desired diametral pitch.
Source: Ref. 1.

TABLE 5.26 PGT System — Tooth Form for Small Numbers of Teeth
(PGT-4)[a]

Number of teeth, n	Minimum circular tooth thickness, t	Outside diameter, D_O	Root diameter, D_R
6	2.4195	9.3236	4.9718
7	2.3769	10.3580	5.8547
8	2.3343	11.3795	6.6377
9	2.2917	12.3907	7.5936
10	2.2491	13.3926	8.5036
11	2.2065	14.3867	9.3826
12	2.1639	15.3737	10.2695
13	2.1213	16.3545	11.1525
14	2.0787	17.3296	12.0354
15	2.0361	18.2996	12.9184
16	1.9935	19.2650	13.8013
17	1.9509	20.2260	14.6843
18	1.9083	21.1830	15.5673
19	1.8657	22.1363	16.4502
20	1.8231	23.0861	17.3332
21	1.7805	24.0326	18.2161
22	1.7379	24.9760	19.0991
23	1.6953	25.9164	19.9820
24	1.6527	26.8541	20.8650
25	1.6101	27.7891	21.7479
26	1.5708	28.7000	22.6400

[a]Values are for 1.0 diametral pitch. For other values, divide by desired diametral pitch.
Source: Ref. 1.

TABLE 5.27 PGT System — Helical Outside Diameter

Tooth form	Outside diameter
PGT-1	$D_O = \dfrac{1}{P_n}\left(\dfrac{N}{\cos\psi} - 2.3158\right) + (2.7475 \times t_n)$
PGT-2	$D_O = \dfrac{1}{P_n}\left(\dfrac{N}{\cos\psi} - 2.0158\right) + (2.7475 \times t_n)$

Source: Ref. 1.

TABLE 5.28 PGT System — Helical Root Diameter

Tooth form	Root diameter
PGT-1	$D_R = \dfrac{1}{P_n}\left(\dfrac{N}{\cos\psi} - 6.9758\right) + (2.7475 \times t_n)$
PGT-2	$D_R = \dfrac{1}{P_n}\left(\dfrac{N}{\cos\psi} - 7.2758\right) + (2.7475 \times t_n)$

D_O = outside diameter; D_R = root diameter; P_n = normal diametral pitch; N = number of teeth; ψ = helix angle; t_n = normal circular tooth thickness on standard pitch circle.
Source: Ref. 1.

When the pinion and gear both have 35 teeth or more:

$$t_1 = \left(\frac{N_1 \times (N_2 - 2.0938)}{N_1 - 2.0938}\right)\left(\frac{t_2}{N_2} + \frac{0.01490438 - \text{inv }\phi_2}{P}\right)$$

$$- N_1\left(\frac{0.01490438 - \text{inv }\phi_1}{P}\right) \tag{5.19}$$

TABLE 5.29 PGT System — Helical Gearing Minimum Normal
Circular Tooth Thickness

Tooth form	Minimum normal circular tooth thickness
PGT-1	$t_n = \dfrac{1}{P_n} \left(2.3329 - \dfrac{N \times (1 - \cos^2\phi)}{2.7475 \times \cos \psi} \right)$
PGT-2	$t_n = \dfrac{1}{P_n} \left(2.4793 - \dfrac{N \times (1 - \cos^2\phi)}{2.7475 \times \cos \psi} \right)$

t_n = minimum normal circular tooth thickness; P_n = normal
diametral pitch; N = number of teeth; ψ = helix angle; ϕ =
$\tan^{-1} \left(\dfrac{0.36397023}{\cos \psi} \right)$.
Source: Ref. 1.

where

t_1 = circular tooth thickness of
pinion
t_2 = circular tooth thickness of
gear
N_1 = number of teeth in pinion
N_2 = number of teeth in gear
P = diametral pitch

$$\phi_1 = \cos^{-1} \left(\frac{0.93969262 \times N_1}{N_1 - 2.0938} \right) \quad (5.20)$$

$$\phi_2 = \cos^{-1} \left(\frac{0.93969262 \times N_2}{N_2 - 2.0938} \right) \quad (5.21)$$

Figure 5.25 illustrates the concept of equal strength teeth. The
critical stress area at the root radius is equal in thickness for the
13-tooth pinion and the 60-tooth gear.

Helical gears of the PGT system are modified according to Tables
5.27 to 5.29 and the procedure for calculating normal circular tooth
thicknesses for balanced tooth strength, as expressed below.

Pinion and gear both having fewer teeth than given by

$$N = \frac{2.0938 \times \cos \psi}{1 - \cos \phi}$$

$$t_{n1} = \frac{1}{P_n}\left[2.3329 - \frac{.36397023 \times N_1 \times (1 - \cos \phi)}{\cos \psi}\right] \qquad \text{(a)}$$

$$t_{n2} = \frac{1}{P_n}\left[2.3329 - \frac{.36397023 \times N_2 \times (1 - \cos \phi)}{\cos \psi}\right] \qquad \text{(b)}$$

Pinion having fewer teeth and gear having more teeth than given by

$$N = \frac{2.0938 \times \cos \psi}{1 - \cos \phi}$$

$$t_{n1} = \frac{1}{P_n}\left[2.3329 - \frac{.36397023 \times N_1 \times (1 - \cos \phi)}{\cos \psi}\right] \qquad \text{(a)}$$

$$t_{n2} = \frac{N_2}{P_n}\left[\frac{P_n \times B \times \cos \psi}{N_2 - (2.0938 \times \cos \psi)} + \text{inv } \phi_2 - \text{inv } \phi\right] \qquad \text{(b)}$$

Pinion and gear both having more teeth than given by

$$N = \frac{2.0938 \times \cos \psi}{1 - \cos \phi}$$

$$t_{n1} = \left[\frac{N_1 \times (N_2 - (2.0938 \times \cos \psi))}{N_1 - (2.0938 \times \cos \psi)}\right]\left[\frac{t_{n2}}{N_2} + \frac{\text{inv } \phi - \text{inv } \phi_2}{P_n}\right]$$

$$\qquad - N_1\left[\frac{\text{inv } \phi - \text{inv } \phi_1}{P_n}\right]$$

where

N = number of teeth determining which equation to use
ψ = helix angle
ϕ = $\tan^{-1}\left(\dfrac{0.36397023}{\cos \psi}\right)$
N_1 = number of teeth in pinion
N_2 = number of teeth in gear
t_{n1} = normal circular tooth thickness of pinion
t_{n2} = normal circular tooth thickness of gear
ϕ_1 = $\cos^{-1}\left[\dfrac{N_1 \times \cos \phi}{N_1 - (2.0938 \times \cos \psi)}\right]$
ϕ_2 = $\cos^{-1}\left[\dfrac{N_2 \times \cos \phi}{N_2 - (2.0938 \times \cos \psi)}\right]$
B = $\dfrac{N_1 \times \cos \phi}{P_n \times \cos \psi}\left(\dfrac{P_n \times t_{n1}}{N_1} + \text{inv } \phi\right)$

The PGT system is not standard in the gearing industry. Therefore, any specification of gears made under the PGT system must provide the mold manufacturer, molder, and inspector with the proper
basic rack tooth form as given in any of Figures 5.21, 5.22, 5.23, or
5.24. This is an important requirement without which the gears cannot be made to specification.

References

1. *Plastics Gearing*, ABA/PGT Publishing Company, Manchester,
   CT (1976).

2. R. E. Peterson, *Stress Concentration Factors*, John Wiley &
   Sons, NY (1974). Original material by T. J. Dolan, E. L.
   Broghamer, University of Illinois Engineering Experiment Sta-
   tion Bulletin 335 (1942), used with permission from University
   of Illinois, Urbana, IL.

3. *Design and Production of Gears in Celcon Acetal Copolymer*,
   Celanese Plastics Company, Newark, NJ (1969).

4. *Sier-Bath Precision Gears/Gear Systems*, Sier-Bath Gear Co.,
   North Bergen, NJ (1968).

5. G. W. Michalec, *Precision Gearing Theory and Practice*, John
   Wiley & Sons, NY (1966).

6. *Design Manual for Fine-Pitch Gearing* (AGMA 370.01), American
   Gear Manufacturers Association, Alexandria, VA (1973).

7. *Tooth Proportions for Coarse-Pitch Involute Spur Gears*, USAS
   B6.1-1968 (AGMA 201.02), American Gear Manufacturers Association, Alexandria, VA (1974).

8. E. Buckingham, *Analytical Mechanics of Gears*, Dover Publications, NY (1949).

9. A. H. Candee, *Introduction to the Kinematic Geometry of Gear
   Teeth*, Chilton Company, NY (1961).

10. H. E. Merritt, *Gears*, Sir Isaac Pitman and Sons, London
    (1955).

11. W. A. Tuplin, *Gear Design*, Industrial Press, NY (1962).

12. L. D. Martin, *Tooling for Injection Molded Nylon Gears*, The
    Tool Engineer 33:1 (July 1954), pp. 61-66.

13. L. D. Martin "The A.B.C.'s of Designing Injection Molded Nylon
    Gears," *Machinery* 62:3 (November 1955, pp. 151-156, and
    December 1955, pp. 162-165).

14. E. Miller, Ed., *Plastics Products Design Handbook, Part A, Materials and Components*, Marcel Dekker, NY (1981).

15. *Fine- and Coarse-Pitch Hobs* (AGMA 120.01), American Gear Manufacturers Association, Alexandria, VA (1975).

16. *Concurve*® *High Performance Gearing By Spiroid*, U.S. Patent 3,631,736, Illinois Tool Works, Chicago, IL (1981).

17. G. L. Breur, *Proposed Revision of Tooth Proportions for Enlarged Pinions* (AGMA Paper 209.10), American Gear Manufacturers Association, Alexandria, VA (1971).

18. *AGMA Standard System Tooth Proportions for Fine-Pitch Involute Spur and Helical Gears* (AGMA 207.06-1977 and ANSI B6.7-1977), American Gear Manufacturers Association, Alexandria, VA (1977).

6

Gear Rating

Gear rating is the mathematical procedure for estimating the load-carrying capacity of gears. It is usually made as early in the design process as possible, when all the application data is gathered and anticipated load calculations compared with the rating of the gear-set. The gear-set load rating must always be equal to, or greater than, the anticipated application loading. Equal rating and application loads are acceptable if conservative values are used for individual factors, peak loads are accounted for, or an adequate safety factor is used.

Size Estimation

Most authorities recommend that a preliminary sizing be made to determine feasibility of the gear design (1,2,3,4). For acetal and nylon, Figures 6.1 through 6.5 provide a quick means of gear strength estimation.

Figure 6.1 can be used to determine either diametral pitch, tangential tooth load at the pitch circle, face width, or allowable stress if three parameters are known, as illustrated on the nomograph. A combination of stress and load projected to the reference line becomes a fulcrum point for selection of a combination of diametral pitch and face width. If three parameters are known, the fourth parameter is fixed by the nomograph. The familiar Lewis equation has been simplified and conservative values of design stress and a tooth form factor of 0.6 is used.

Figures 6.2 and 6.3 will allow the adjustment of stress where lubrication and temperatures other than 73°F must be considered. It can be seen that design stress must be reduced if initial rather than

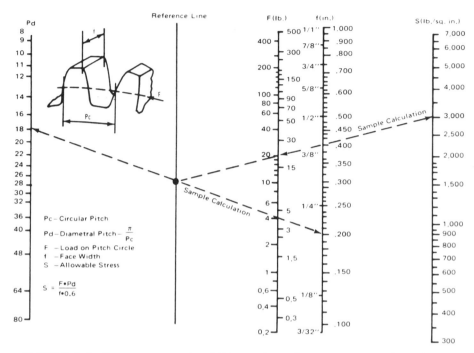

FIGURE 6.1 Gear nomograph, 73°F, estimated gear strength in British units. (From Ref. 4.)

continuous lubrication is provided. The temperature correction is 1.0 at 73°F, 1.1 at 32°F, and 0.8 at 113°F.

Figures 6.4 and 6.5 are given for use in the design of gearing using the module (SI units) system. Figure 6.4 is the preferred system because newtons (N) are the recognized units for tooth loading, Kilograms force can be converted to newtons by multiplying the factor 9.80665. The common practice is to round to a factor of 9.8. Module (m in the SI system) is the pitch diameter in millimeters divided by the number of teeth, is expressed as m = D/N, and relates to diametral pitch as m = 25.4/P.

Earlier, recommended gear tooth size estimates were made using

$$C^2 F = \frac{31,500}{K} \frac{HP}{n_P} \frac{(m_G + 1)^3}{m_G}$$ (6.1)

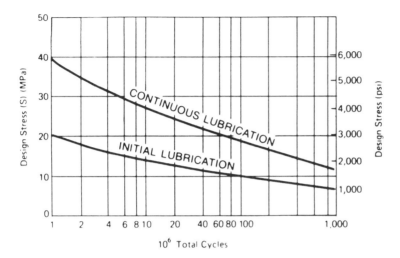

FIGURE 6.2 Suggested design stress for gears. (From Ref. 4.)

where

$$K = \frac{W}{Fd} \frac{m_G + 1}{m_G} \qquad (6.2)$$

C = center distance, in.
F = net face width, in.
HP = horsepower
n_P = pinion, rpm
m_G = ratio = n_G/n_P
W = tangential driving load, lb = $2T_P/d$ (6.3)
T_P = pinion torque, lb-in.
d = pinion pitch diameter, in.

For plastics, the value of K was typically 50 or less for a gear run against a hardened steel pinion with uniform loading and pitch-line velocity under 1000 fpm.

As materials have developed and molded parts have become more uniform, several means of rating have evolved. The general conclusion, however, is that the basic formulas for rating metal gears apply to the rating of plastics gears. A prudent choice of stress level is an absolute necessity when using plastics. In a sense, this correlation is a vindication of the earlier studies indicating that gears fail

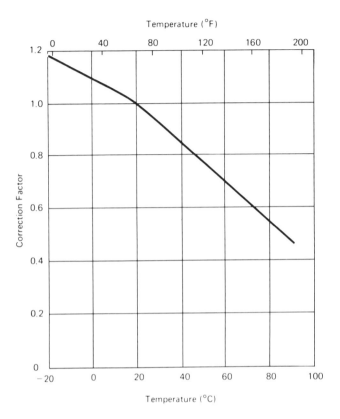

FIGURE 6.3 Correction of design stress for different operating temperatures. (From Ref. 4.)

in basically the same way regardless of material. One point to note,
however, is that although plastics gears can be run without lubrication, a proper rating should be made considering the presence or lack
of lubrication, the type of lubrication, and the method used. As discussed elsewhere, the lubrication aspect is directly related to the temperature generated at the gear mesh. Unlubricated gears usually fail
due to wearing of the tooth profile, pitting of the contact surface, or
melting of the tooth. Several references (1,2,5) give a concise description of surface stress reactions due to the combined effect of
rolling and sliding in the tooth contact regions. Lubricated gears
usually fail at the tooth root due to bending fatigue.

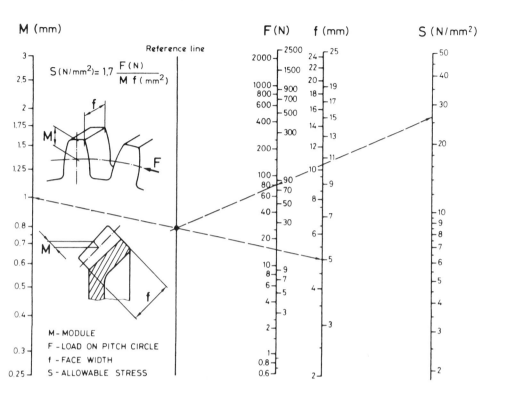

FIGURE 6.4 Gear nomograph (SI units-newtons). (From Ref. 3.)

Tooth Strength with Pitch-Line Loading

Studies have been made comparing the several methods of rating plastics gearing. If the application is critical, the materials chosen by the size estimation method should be evaluated further using the Lewis equation of the form:

$$S_b = \frac{FP_d}{fY} \qquad (6.4)$$

where

S_b = bending stress
F = tangential tooth loading at the pitch line

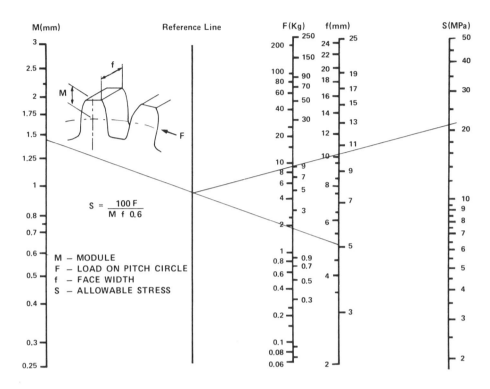

FIGURE 6.5 Gear nomograph, 23°C (SI units-kilograms). (From Ref. 4.)

TABLE 6.1 Lewis Form Factor Y for Loading near the Pitch Point[a]

Number of teeth	14 1/2°	20° full depth	20° stub
12	0.356	0.415	0.502
13	0.373	0.442	0.524
14	0.383	0.468	0.540
15	0.400	0.490	0.565
16	0.410	0.500	0.577

(continued)

TABLE 6.1 (continued)

Number of teeth	14 1/2°	20° full depth	20° stub
17	0.425	0.512	0.588
18	0.442	0.520	0.605
19	0.463	0.533	0.617
20	0.480	0.544	0.626
21	0.490	0.551	0.640
22	0.495	0.557	0.646
24	0.506	0.571	0.665
26	0.521	0.587	0.677
28	0.533	0.595	0.687
30	0.541	0.605	0.697
34	0.555	0.629	0.712
38	0.564	0.650	0.730
43	0.574	0.671	0.738
50	0.589	0.696	0.756
60	0.602	0.712	0.775
75	0.614	0.734	0.791
100	0.621	0.758	0.807
150	0.636	0.780	0.832
300	0.650	0.802	0.854
Rack	0.660	0.824	0.882

[a]Lewis form factor Y is used when the tangential tooth loading is calculated at the pitch point and form factor y of Table 6.3 is used for loading at the tooth tip. When using values from Table 6.1, design stress values can be taken directly from materials property tables. When using values from Table 6.3, reduced stress values should be used such as the maximum allowable stress values developed for the specific material.
Source: Ref. 5.

TABLE 6.2 Safe Stress

Plastic	Safe stress, psi	
	Unfilled	Glass-reinforced
ABS	3000	6000
Acetal	5000	7000
Nylon	6000	12000
Polycarbonate	6000	9000
Polyester	3500	8000
Polyurethane	2500	

Source: Ref. 6.

P_d = diametral pitch
f   = face width
Y = form factor from Table 6.1 (see note)

Tests have also shown that the most severe tooth loading occurs when the tangential load is at the pitch line and the number of pairs of teeth in mesh approaches 1.0. The tooth form factor (Y) of Table 6.1 is given for loading at the pitch line since tooth flexure generally always provides more than one set of teeth in contact. In the Lewis equation, torque (T) in in./lb can be related to horsepower (HP) and revolutions per minute (n) by

$$T = \frac{63,025 \text{ HP}}{n} \qquad (6.5)$$

and pitch diameter (D) and diametral pitch (P) by

$$T = \frac{\sigma f Y D}{2 P} \qquad (6.6)$$

Therefore,

$$HP = \frac{\sigma f Y D n}{126,050 P} \qquad (6.7)$$

Safe stress (σ) is specified from Table 6.2 that allows for moderate temperature increase and some initial lubrication. Pitch line velocity

TABLE 6.3 Tooth Form Factor y for Loading near the Tooth Tip[a]

Number of teeth	14 1/2° involute or cycloidal	20° full depth involute	20° stub tooth involute	20° internal full depth Pinion	Gear
12	0.210	0.245	0.311	0.327	-
13	0.220	0.261	0.324	0.327	-
14	0.226	0.276	0.339	0.330	-
15	0.236	0.289	0.348	0.330	-
16	0.242	0.295	0.361	0.333	-
17	0.251	0.302	0.367	0.342	-
18	0.261	0.308	0.377	0.349	-
19	0.273	0.314	0.386	0.358	-
20	0.283	0.320	0.393	0.364	-
21	0.289	0.327	0.399	0.371	-
22	0.292	0.330	0.405	0.374	-
24	0.298	0.336	0.415	0.383	-
26	0.307	0.346	0.424	0.393	-
28	0.314	0.352	0.430	0.399	0.691
30	0.320	0.358	0.437	0.405	0.679
34	0.327	0.371	0.446	0.415	0.660
38	0.336	0.383	0.456	0.424	0.644
43	0.346	0.396	0.462	0.430	0.628
50	0.352	0.408	0.474	0.437	0.613
60	0.358	0.421	0.484	0.446	0.597
75	0.364	0.434	0.496	0.452	0.581
100	0.371	0.446	0.506	0.462	0.565
150	0.377	0.459	0.518	0.468	0.550
300	0.383	0.471	0.534	0.478	0.534
Rack	0.390	0.484	0.550	-	-

[a]See note under Table 6.1.
Source: Ref. 6.

TABLE 6.4 Service Factor, C_S

Type of load	8--10 hours per day	24 hours per day	Intermittent (3 hours per day)	Occasional (1/2 hour per day)
Steady	1.00	1.25	0.80	0.50
Light shock	1.25	1.50	1.00	0.80
Medium shock	1.50	1.75	1.25	1.00
Heavy shock	1.75	2.00	1.50	1.25

Source: Ref. 6.

(V) in fpm is factored into the calculations when test data have been taken at an extreme value from that of the application and are as follows.

$$HP = \frac{\sigma\ f\ y\ V}{55(600+V)P\ C_S} \tag{6.8}$$

In this case the loading is taken at the tooth tip using (y) from Table 6.3 and service factor (C_S) from Table 6.4. Reference (6) specifies equation 6.9 for helical gearing and equation 6.10 for straight bevel gearing. Tooth form factors continue to be taken at the tooth tip; P_n is normal diametral pitch and C is the pitch cone radius, referred to as outer cone distance in Figure 6.6 and Table 6.6.

For helical gearing:

$$HP = \frac{\sigma\ F\ y\ V}{423(78 + \sqrt{V})P_n\ C_S} \tag{6.9}$$

$$P_n = \frac{P_d}{\cos\ \text{helix angle}} \tag{6.10}$$

For bevel gearing:

$$HP = \frac{\sigma\ F\ y\ V\ (C\text{-}F)}{55(600+V)PC\ C_S} \tag{6.11}$$

$$C = \frac{D}{2\ \sin\ \Gamma} \tag{6.12}$$

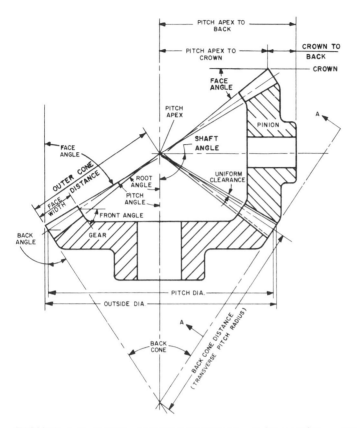

FIGURE 6.6 Bevel gear nomenclature, axial section. (From Ref. 7.)

where

$$D = \frac{N}{P}$$

$$\Gamma = \tan^{-1*}\left(\frac{N_g}{N_p}\right)$$

= gear pitch angle

$*\tan^{-1}$ (b/A) = arctangent b/A = angle whose tangent is b/A.

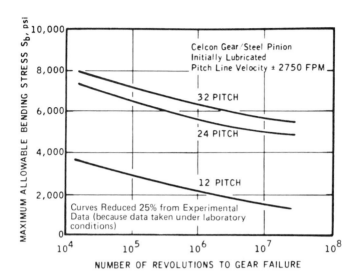

FIGURE 6.7 Bending stress, S_b, vs. fatigue cycles. (From Ref. 5.)

In equations (6.4) through (6.12), a design factor K has not been used. This is because stress levels are not taken at maximum values and compensation for variables is accounted for in the values of Tables 6.2 through 6.4.

Where actual test data has been developed, it is possible to modify allowable bending stress using specific factors to arrive at safe bending stress. For example, reference (5) gives

$$S_b \text{ (allowable)} = S_b \text{ (graph) } K_T K_L K_M \tag{6.13}$$

where

K_T = temperature factor
K_L = lubrication factor
K_M = material factor

The temperature factor (K_T) is calculated as the tensile strength of the material at the operating temperature divided by the tensile strength of the material at the temperature at which the data were taken.

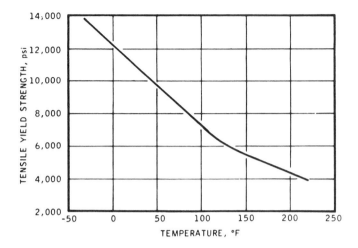

FIGURE 6.8 Tensile strength vs. temperature for Celcon acetal co-polymer. (From Ref. 5.)

Continuous lubrication increases the allowable bending stress by a factor of at least 1.5 with a range of 1.5 to 3.0. $K_L = 1.0$ for the initial lubrication case. This illustrates the importance of lubrication even on plastics gears.

Mating material factor (K_M) is given as 1.0 for a Celcon® gear with a steel pinion of roughness 32 microinches (rms) and 0.75 for a Celcon gear with a Celcon pinion.

Figure 6.7 provides S_b (allowable values) and Figure 6.8 tensile strength vs. temperature values for the K_T factor.

Design factor (K) can be used with equations (6.6) and (6.7) with modifications such as

$$T = \frac{S_b D f Y K}{2P} \tag{6.14}$$

$$HP = \frac{S_b D f Y n K}{126,050P} \tag{6.15}$$

Table 6.1 and Figures 6.7 and 6.8 are used because data are specified at the pitch line and bending stress and various operating conditions have been determined by actual testing.

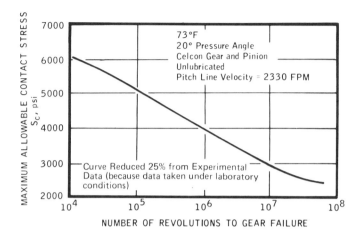

FIGURE 6.9   Surface contact stress, S_c, vs. fatigue cycles.   (From Ref. 5.)

It is recommended that after determination of the material to be used, the material supplier be consulted for any recent test data available for use in strength calculations.

Wear Failure

Because unlubricated plastics gears tend to fail by wear rather than by fatigue, contact stress must be the design criterion.   Calculation of surface contact stress (using Hertzian stress) for the application is

$$S_c = \sqrt{\frac{W}{f\ Dp_p} \frac{(r+1)}{r}} \times \sqrt{\frac{0.7\ E_1\ E_2}{(E_1 + E_2)\ \cos\ \phi\ \sin\ \phi}} \qquad (6.16)$$

where

W = tangential driving load, lb = $2\ T_p/d$
Tp = pinion torque, lb-in.
d = pinion pitch diameter, in.
f = face width
Dp_p = pinion pitch diameter

r = gear ratio = N_g/N_p
E_1, E_2 = moduli of elasticity of mating gear materials
ϕ = pressure angle

The application contact stress of equation (6.16) must be less than allowable contact stress values after being adjusted for operating temperature, material combinations, or pressure angles that differ from those conditions under which test data was developed. It may be a problem to obtain contact stress values that adequately apply to gearing. If the data is not readily available, consultation should be made with the material supplier and agreement reached as to the correct values to use.

Reference 5 gives contact stress, test conditions used to develop the number of fatigue cycles, and

$$S_c \text{ (allowable)} = S_c \text{ (graph)} \times K_J \tag{6.17}$$

where S_c (graph) is selected from Figure 6.9.

Factor (K_J) is $\dfrac{C_K}{C_{K_s}}$

C_K is the Hertz stress factor calculated using data from equation (6.18) and (6.18a) and Figures 6.10 and 6.11.

$$C_k = \sqrt{\frac{0.70}{\left(\dfrac{1}{E_1} + \dfrac{1}{E_2}\right)\cos\phi}} \tag{6.18}$$

Value of flexural modulus to 250°F for steel = 29 X 10^6 (6.18a)
psi, constant over temperature range (-50 to +250°F)

C_{K_s} is the Hertz stress factor for conditions at which data were taken. Therefore

$$K_J = \frac{C_K}{C_{K_s}}$$

$$K_J = \frac{C_K}{639}$$

Once again, the application contact stress must be less than the allowable contact stress.

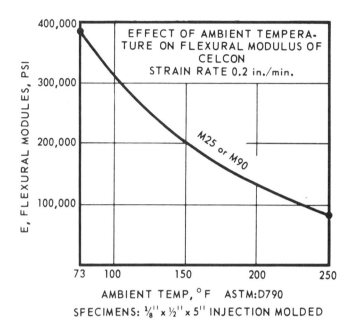

FIGURE 6.10 Values of flexural modulus to 250°F, for Celcon. (From Ref. 5.)

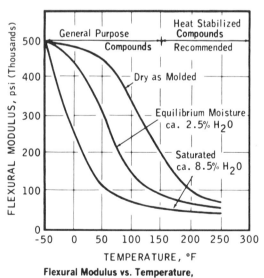

FIGURE 6.11 Values of flexural modulus to 250°F, for Nylon 6,6. (From Ref. 5.)

TABLE 6.5 Basic Dimensions for Helical Gears

P_d = diametral pitch in plane of rotation

P_{d_n} = diametral pitch in normal profile plane

ϕ = pressure angle in plane of rotation

ϕ_n = pressure angle in normal profile plane

a = addendum

h = whole depth

G = groove width

D_o = outside diameter

d = dedendum

C = center distance

f = face width

N = number of teeth

D_p = pitch diameter

α = helix angle, degress

C_1 = clearance

f_a = active face

$$P_d = \frac{N}{D_p}$$

$$P_{d_n} = \frac{P_d}{\cos \alpha}$$

$$C_1 = \frac{0.157}{P_d} \text{ to } \frac{0.30}{P_d}$$

$$h = 2a + C_1$$

$$f_a = \frac{7.22568}{P_d \times \tan \alpha}$$

$$C \text{ (external)} = \frac{D_{p_G} + D_{p_P}}{2}$$

$$\tan \phi = \frac{\tan \phi_n}{\cos \alpha}$$

$$a = \frac{0.7}{P_d} \text{ to } \frac{1.0}{P_d}$$

$$d = a + C_1$$

$$D_o = \frac{N}{P_d} + 2a$$

$$f = f_a + G$$

$$C \text{ (internal)} = \frac{D_{p_G} - D_{p_P}}{2}$$

Source: Ref. 5.

Helical Gear Rating

The rating of helical gears follows the same basic method as for spur gears. However, it can be seen in equations (6.19) and (6.20) that diametral pitch (P_d) and the Lewis form factor (Y) are selected based on the formative number of teeth calculated by equation (6.21). The term for helix angle (α) is used to calculate the formative number of

TABLE 6.6 Example Straight Bevel Gear-Set Calculations (all linear dimensions in inches)

#	Description	Symbol / Formula	Value
1	Number of pinion teeth	n	16
2	Number of gear teeth	N	49
3	Diametral pitch	P_d	5
4	Face width	F	1.5
5	Working depth	$h_K = \dfrac{2.000}{P_d}$	0.400
6	Whole depth	$h_t = \dfrac{2.188}{P_d} + .002$	0.440
7	Pressure angle	ϕ	20°
8	Shaft angle	Σ	90°

#	Description	Pinion formula	Pinion value	Gear formula	Gear value
9	Pitch diameter	$d = \dfrac{n}{P_d}$	3.2000	$D = \dfrac{N}{P_d}$	9.8000
10	Pitch angle	$\gamma = \tan^{-1}\dfrac{\sin \Sigma}{N/n + \cos \Sigma^{a}}$	18°5'	$\Gamma = \Sigma - \gamma$	71°55'
11	Outer cone distance	$A_o = \dfrac{D}{2 \sin \Gamma}$	5.1546		
12	Circular pitch	$p = \dfrac{3.1416}{P_d}$	0.6283		
13	Right angle gear ratio	$m_{90} = \sqrt{\dfrac{N \cos \gamma}{n \cos \Gamma}}$	3.0625		
14	Outer addendum	$a_{oP} = h_K - a_{oG}$	0.282	$a_{oG} = \dfrac{0.540}{P_d} + \dfrac{0.460}{P_d m_{90}^2}$	0.118

No.	Description	Pinion	Value	Gear	Value
15	Outer dedendum	$b_{oP} = h_t - a_{oP}$	0.157	$b_{oG} = h_t - a_{oG}$	0.322
16	Clearance	$c = h_t - h_K$	0.040		
17	Dedendum angle	$\delta_P = \tan^{-1}\dfrac{b_{oP}}{A_o}$	1°44'	$\delta_G = \tan^{-1}\dfrac{b_{oG}}{A_o}$	3°33'
18	Face angle of blank	$\gamma_o = \gamma + \delta_G$	21°38'	$\Gamma_o = \Gamma + \delta_P$	73°39'
19	Root angle	$\gamma_R = \gamma - \delta_P$	16°21'	$\Gamma_R = \Gamma - \delta_G$	68°22'
20	Outside diam.	$d_o = d + 2a_{oP}\cos\gamma$	3.736	$D_o = D + 2a_{oG}\cos\Gamma$	9.873
21	Pitch apex to crown	$x_o = A_o\cos\gamma - a_{oP}\sin\gamma$	4.813	$X_o = A_o\cos\Gamma - a_{oG}\sin\Gamma$	1.488
22	Outer circular thickness	$t = p - T$	0.3814	$T = \dfrac{P}{2} - (a_{oP} - a_{oG})\tan\phi - \dfrac{K}{P_d}$	0.247
23	Normal backlash	$B = $ (Table 8.5, Quality No. Q10)	0.006		
24	Outer chordal thickness	$t_c = t - \dfrac{t^3}{6d^2} - \dfrac{B}{2\cos\phi}$	0.378	$T_C = T - \dfrac{T^3}{6D^2} - \dfrac{B}{2\cos\phi}$	0.245
25	Outer chordal addendum	$a_{cP} = a_{oP} + \dfrac{t^2\cos\gamma}{4d}$	0.293	$a_{cG} = a_{oG} + \dfrac{T^2\cos\Gamma}{4D}$	0.118

[a]Note that cosine Σ is negative if $\Sigma > 90°$.
Source: Ref. 7.

teeth and compressive stress. Equations (6.19) and (6.20) are application stress levels that are compared to, and must be less than, allowable stress. Basic dimensions are provided in Table 6.5 for helical gears and F_T is tangential force in pounds.

$$S_b = \frac{F_T P_d n Cs}{f Y} \tag{6.19}$$

$$S_c = \sqrt{\frac{\cos^2 \alpha_p F_T}{f Dp_p} \left(\frac{N'_G + N'_p}{N'_g} \right)} \tag{6.20}$$

$$N' = \frac{N}{\cos^3 \alpha} \tag{6.21}$$

Bevel Gear Rating

Rating of straight bevel gears also follows closely the rating methods
used for spur gears. In this case, the formative number of teeth (N')
is calculated using the pitch angle (μ) as

$$N' = N/\cos \mu \tag{6.22}$$

Bending stress is calculated by

$$S_b = \frac{F_T P_d L_p C_s}{f Y (L_p - f)} \tag{6.23}$$

where L_p is outer cone distance or pitch cone radius, and F_T is tangential force in pounds or $2T/D_p$ and contact stress is

$$S_c = \sqrt{\frac{(\cos \mu_p) F_T}{D_{o_p} f} \left(\frac{N'_p + N'_g}{N'_g} \right)} \times C_K \tag{6.24}$$

Table 6.6 is an example of a straight bevel gear-set calculation sheet
(7). For the above bending and contact stress calculations pitch angle,
face width, pitch cone radius, and outside diameter must be known.
Note that outside diameter instead of pitch diameter is to be used in
equations calculating torque or horsepower.

The example shown in Table 6.6 can be used as a design example
using either Figures 6.12, 6.13, or 6.14, depending on the required
pressure angle. All linear dimensions are in inches. Table 6.7 is the
same design example but with linear dimensions in millimeters.

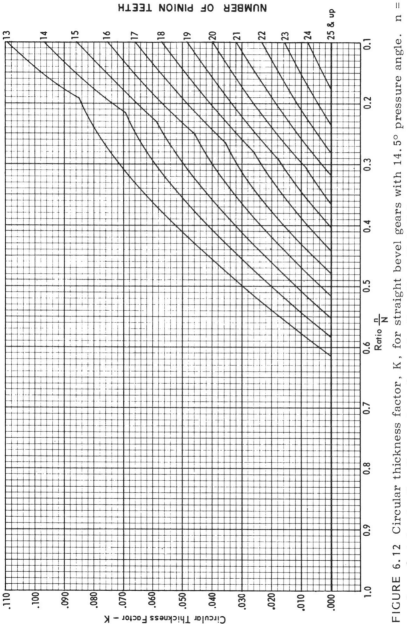

FIGURE 6.12 Circular thickness factor, K, for straight bevel gears with 14.5° pressure angle. n = number of pinion teeth. N = number of gear teeth. (From Ref. 7.)

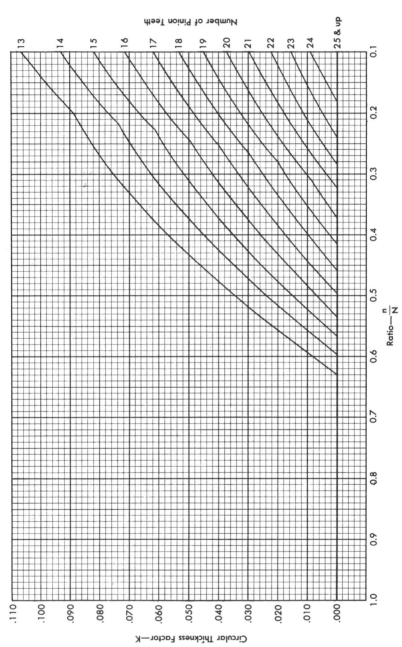

FIGURE 6.13 Circular thickness factor, K, for straight bevel gears with 20° pressure angle. n = number of pinion teeth. N = number of gear teeth. (From Ref. 7.)

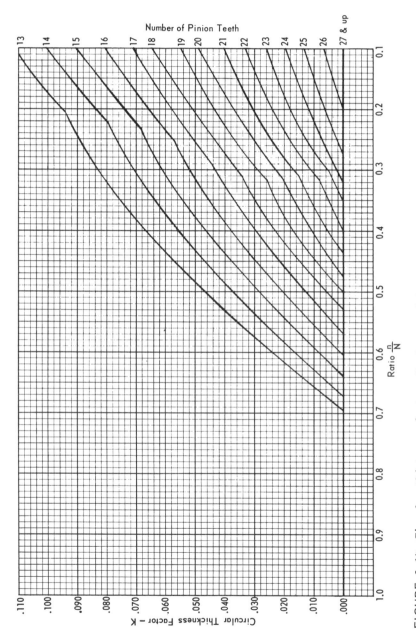

FIGURE 6.14 Circular thickness factor, K, for straight bevel gears with 25° pressure angle. n = number of pinion teeth. N = number of gear teeth. (From Ref. 7.)

TABLE 6.7 Example Straight Bevel Gear-Set Calculations (all linear dimensions in mm)

1	Number of pinion teeth	n	16	5	Working depth $h_K = 2.00\,m$	10.16
2	Number of gear teeth	N	49	6	Whole depth $h_t = 2.188 + 0.05$	11.17
3	Module	m	5.08	7	Pressure angle ϕ	20°
4	Face width	F	38.1	8	Shaft angle Σ	90°

		Pinion		Gear	
9	Pitch diameter	$d = nm$	81.28	$D = Nm$	248.92
10	Pitch angle	$\gamma = \tan^{-1}\dfrac{\sin \Sigma}{N/n + \cos \Sigma}$	18°5'	$\Gamma = \Sigma - \gamma$	71°55'
11	Outer distance	$A_o = \dfrac{D}{2 \sin \Gamma}$	130.93		
12	Circular pitch	$p = 3.1416\,m$	15.96		
13	Right angle gear ratio	$m_{90} = \sqrt{\dfrac{N \cos \gamma}{n \cos \Gamma}}$	3.0625		
14	Outer addendum	$a_{oP} = h_K - a_{oG}$	7.17	$a_{oG} = 0.540\,m + \dfrac{0.460\,m}{m_{90}^2}$	2.99
15	Outer dedendum	$b_{oP} = h_t - a_{oP}$	4.0	$b_{oG} = h_t - a_{oG}$	8.17
16	Clearance	$c = h_t - h_K$	1.01		

		Pinion		Gear
17	Dedendum angle	$\delta_P = \tan^{-1} \dfrac{b_{oP}}{A_o}$	1°44'	$\delta_G = \tan^{-1} \dfrac{b_{oG}}{A_o}$ 3°33'
18	Face angle of blank	$\gamma_o = \gamma + \delta_G$	21°38'	$\Gamma_o = \Gamma + \delta_P$ 73°39'
19	Root angle	$\gamma_R = \gamma - \delta_P$	61°21'	$\Gamma_R = \Gamma - \delta_G$ 68°22'
20	Outside diameter	$d_o = d + 2a_{oP} \cos \gamma$	94.91	$D_o = D + 2a_{oG} \cos \Gamma$ 250.78
21	Pitch apex to crown	$x_o = A_o \cos \gamma - a_{oP} \sin \gamma$	122.24	$X_o = A_o \cos \Gamma - a_{oG} \sin \Gamma$ 37.80
22	Outer circular thickness	$t = p - T$	9.69	$T = \dfrac{P}{2} - (a_{oP} - a_{oG}) \tan \phi - K_m$ 6.27
23	Normal backlash	$B = $ (Table 8.5, Quality No. Q10) 0.15		
24	Outer chordal thickness	$t_c = t - \dfrac{t^3}{6d^2} - \dfrac{B}{2 \cos \phi}$	9.60	$T_C = T - \dfrac{T^3}{6D^2} - \dfrac{B}{2 \cos \phi}$ 6.22
25	Outer chordal addendum	$a_{cP} = a_{oP} + \dfrac{t^2 \cos \gamma}{4d}$	7.44	$a_{cG} = a_{oG} + \dfrac{T^2 \cos \Gamma}{4D}$ 3.00

[a]Note that cosine Σ is negative if $\Sigma > 90°$.
Source: Ref. 7.

In the above discussion of gear rating, the data will be influenced by the use, or lack of, tooth tip modifications and full round root radius as given in Chapter 5 on tooth modification.

References

1. D. W. Dudley, *Practical Gear Design*, McGraw-Hill, New York (1954).

2. D. W. Dudley, Ed., *Gear Handbook*, McGraw-Hill, New York (1962).

3. *DuPont Delrin Acetal Resin Design Handbook*, E. I. DuPont DeNemours and Co., Wilmington, DE (1981).

4. *DuPont Zytel Nylon Resin Design Handbook*, Bulletin E-44971, E. I. DuPont DeNemours and Co., Wilmington, DE.

5. *Design and Production of Gears in Celcon Acetal Copolymer*, Celanese Plastics Company, Newark, NJ (1979).

6. *Plastics Gearing*, ABA/PGT Publishing Company, Manchester, CT (1976).

7. *AGMA Standard System for Straight Bevel Gears* (AGMA 208.03), American Gear Manufacturers Association, Alexandria, VA (1978).

7
Tolerance

Tables of tolerance values are provided for blanks, the part configuration, tooth conditions, and the assembly. Care should be exercised in applying the values for they are given as guidance for general applications. Because it is impossible to provide exact values for all applications, it may be necessary to make a complete dimensional study to ensure proper tolerancing for specific cases. The study will include thermal expansion and moisture absorption for plastics gears. Where quality class applies, the stated values for that class must be maintained.

It is important to note that values in standards are determined by consensus and compromise to ensure wide acceptance. Usage, then, must often be augmented by more exacting specifications for a particular engineering discipline or company product. This in no sense implies that quality class tolerances can be varied for a particular quality class.

Specifications showing quality class number or letter designations imply that published tolerances for that class should apply. In a case in which specific tolerances also are given, tolerances on the part print are to be used. Specification of both numerical tolerances and the AGMA quality numbers or letters on the same part print are a form of double dimensioning and should be avoided to prevent confusion.

Blanks

When plastics gears are cut, it is as important as it is in metal gears to use adequately prepared gear blanks. Close toleranced gear teeth

TABLE 7.1 AGMA Quality Number vs. Gear Blank Quality

AGMA Quality Number	Normal Diametral Pitch	Tooth-to-Tooth Composite Tolerance		Total Composite Tolerance		Blank Quality No.	
		1" PD	4" PD	1" PD	4" PD	1" PD	4" PD
Q5	20	.00356	.00334	.00743	.00886	K	L
	30	.00308	.00303	.00628	.00767	I	J
	40	.00283	.00283	.00572	.00693	H	J
	80	.00240	.00240	.00457	.00543	G	H
	100	.00227	.00227	.00425	.00502	F	G
Q6	20	.00255	.00239	.00530	.00633	H	K
	30	.00220	.00217	.00449	.00548	G	I
	40	.00202	.00202	.00409	.00495	G	H
	80	.00171	.00171	.00327	.00388	F	G
	100	.00162	.00162	.00304	.00359	E	F
Q7	20	.00182	.00171	.00379	.00452	G	I
	30	.00157	.00155	.00321	.00392	F	H
	40	.00144	.00144	.00292	.00354	F	H
	80	.00122	.00122	.00233	.00277	E	F
	100	.00116	.00116	.00217	.00256	E	F
Q8	20	.00130	.00122	.00271	.00323	G	H
	30	.00112	.00111	.00229	.00280	F	G
	40	.00103	.00103	.00209	.00253	F	G
	80	.00087	.00087	.00167	.00198	E	F
	100	.00083	.00083	.00155	.00183	D	E
Q9	20	.00093	.00087	.00192	.00231	F	G
	30	.00080	.00079	.00164	.00200	E	F
	40	.00074	.00074	.00149	.00180	E	F
	80	.00062	.00062	.00119	.00141	D	E
	100	.00059	.00059	.00111	.00131	C	C
Q10	20	.00066	.00062	.00138	.00165	D	E
	30	.00057	.00056	.00117	.00143	D	E
	40	.00053	.00053	.00106	.00129	D	E
	80	.00045	.00045	.00085	.00101	C	C
	100	.00042	.00042	.00079	.00093	B	B
Q11	20	.00047	.00044	.00099	.00118	C	C
	30	.00041	.00040	.00083	.00102	C	C
	40	.00038	.00038	.00076	.00092	C	C
	80	.00032	.00032	.00061	.00072	B	B
	100	.00030	.00030	.00056	.00067	B	B

(continued)

TABLE 7.1 (continued)

AGMA Quality Number	Normal Diametral Pitch	Tooth-to-Tooth Composite Tolerance		Total Composite Tolerance		Blank Quality No.	
		1" PD	4" PD	1" PD	4" PD	1" PD	4" PD
Q12	20	.00034	.00032	.00070	.00084	C	C
	30	.00029	.00029	.00060	.00073	B	C
	40	.00027	.00027	.00054	.00066	B	C
	80	.00023	.00023	.00043	.00052	B	B
	100	.00022	.00022	.00040	.00048	A	A
Q13	20	.00024	.00023	.00050	.00060	B	C
	30	.00021	.00021	.00043	.00052	B	B
	40	.00019	.00019	.00039	.00047	B	B
	80	.00016	.00016	.00031	.00037	A	B
	100	.00015	.00015	.00029	.00034	A	B
Q14	20	.00017	.00016	.00036	.00043	B	B
	30	.00015	.00015	.00030	.00037	B	B
	40	.00014	.00014	.00028	.00034	B	B
	80	.00012	.00012	.00022	.00026	A	B
	100	.00011	.00011	.00021	.00024	A	A
Q15	20	.00012	.00012	.00026	.00031	A	B
	30	.00011	.00010	.00022	.00027	A	B
	40	.00010	.00010	.00020	.00024	A	A
	80	.00008	.00008	.00016	.00019	A	A
	100	.00008	.00008	.00015	.00017	A	A
Q16	20	.00009	.00008	.00018	.00022	A	A
	30	.00008	.00007	.00016	.00019	A	A
	40	.00007	.00007	.00014	.00017	A	A
	80	.00006	.00006	.00011	.00013	A	A
	100	.00006	.00006	.00011	.00012	A	A

Note: Straight-line interpolation of tolerances may be used for intermediate values of diametral pitch and pitch diameter.
Source: Ref. 2.

cannot be provided without proper attention to the blanks. Table 7.1 indicates the blank quality letter necessary to provide gears to the level indicated. Reference to Table 7.2 under the appropriate letter column and outside diameter tolerances of Table 7.3 will apply. Tolerances in both tables are referenced to the gear blank as shown in Figure 7.1.

TABLE 7.2 Suggested Tolerances for Fine-Pitch Blanks

Blank Feature	Dimension	A[a]	B	C	D	E	F	G	H	I	J	K	L
Bore	Basic Diameter[b]		.00016	.00022	.00031	.00043	.0006	.00083	.0012	.0016	.0022	.0031	.0043
	Percent length to meet tolerance[c]		80%	80%	70%	60%	60%	60%	60%	60%	60%	60%	60%
	Finish, Micro In.		32	32	32	32	63	63	125	125	125	125	125
Journals	Basic Diameter[d]		.00019	.0002	.00028	.00038	To suit bearings						
	Concentricity TIR[e]		.00014	.00019	.00027	.00037	.00052	.00072	.001	.0014	.0019	.0027	.0037
	Finish, Micro In.		32	32	32	63	63	125	125	125	125	125	125
Outside Diameter	Basic Diameter[f]		.001	.0014	.0019	.0027	$+.000 - 20/P_d$						
	Concentricity TIR[g]		.0004	.00056	.00077	.0012	.0015	.0021	.0029	.004	.0056	.0078	.012
	Finish, Micro In.		32	32	63	125	125	125	125	125	125	125	125
Mounting Surface	Convexity[h]		No convexity allowed										
	Lateral Runout — In./In.[i]		.0003	.00036	.00044	.00053	.00064	.00078	.00094	.00114	.00137	.00165	.0020
	Lateral Runout — Max. at any Radius		.00045	.00054	.00066	.0008	.00096	.00125	.0014	.0017	.00205	.0025	.0030
Lateral	Concavity[j]		.0005 per in. for rigid blank; .0003 per in. for flexible blank. Total not to exceed .0015.					.001 per in. for rigid blank; .0005 per in. for flexible blank. Total .003 max.					
	Finish, Micro In.		32	32	63	63	63	125	125	125	125	125	125

aClass "A" normal practice not applicable.

bUnless otherwise specified, this value is the difference between the minimum effective and the maximum dimensional bore size if no qualifying notes are shown, and it is assumed that no part of the bore can exceed these limits.

cThis allows for a certain amount of bell mouth in the bore. If this allowance is shown on the drawing, it will permit the use of reasonably good blanks that might otherwise be rejected.

dThis is based on type of bearing used. In the case of very precise gears, the tolerance and the looseness in the bearing should be limited to appreciably less than the total composite tolerance in the gear.

eIf two cylindrical surfaces are to be used to support the gear, the cylinders should have a common axis. This is more important to the operation of the bearings than to the gears.

fTo avoid fillet interference, the outside diameter tolerance should be made minus relative to the basic or calculated outside diameter.

gIn very fine-pitch teeth, this specification insures sufficient contact ratio in one extreme and lack of interference in the other extreme.

hConvexity permits rocking of the parts.

iIf extreme, wobble of the gear will occur when mounted.

jAlthough undesirable, concavity provides a contact zone at the largest possible diameter and promotes stability of the gear blank.

Source: Ref. 2.

TABLE 7.3 Fine-Pitch Gear Blank Outside
Diameter Tolerance

$(+0.000 - 0.20/P_d)$	
Diametral pitch	Outside diameter tolerance (in.)
20	+0.0 - 0.010
24	+0.0 -0.0083
32	+0.0 -0.0062
48	+0.0 -0.0042
64	+0.0 -0.0031
72	+0.0 -0.0028
96	+0.0 -0.0021
120	+0.0 -0.0017
200	+0.0 -0.001

P_d = diametral pitch.
Source: Ref. 2.

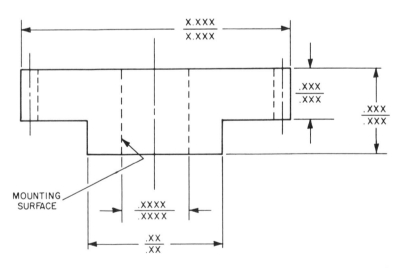

FIGURE 7.1 Gear blank mounting surface example. (From Ref. 2.)

TABLE 7.4 Tolerance on Bore Diameter

AGMA quality number	Tolerance on bore diameter (0 to -, in.)	AGMA quality number	Tolerance on bore diameter (0 to -, in.)
Q5	0.0015	Q11	0.0005
Q6	0.0010	Q12	0.0004
Q7	0.0007	Q13	0.0003
Q8	0.0005	Q14	0.0003
Q9	0.0005	Q15	0.0002
Q10	0.0005	Q16	0.0002

Part Configuration

Table 7.4 contains recommended fine-pitch tolerances on bore diameter using the standard-center-distance system as related to the AGMA quality number. Hub diameters (Table 7.5) were selected to provide adequate strength and length for fastening to the shaft by taper pin, spring pin, or set screw. The adequacy of the mounting method selected should be carefully considered. A review of the material in Chapter 4 may be helpful.

TABLE 7.5 Hub Diameter, Length, and Tolerances

Bore diameter low limit	Hub diameter	Hub length	Taper pin size
0.2500	0.500	0.312	5-0
0.3125	0.625	0.375	4-0
0.3750	0.750	0.438	3-0
0.5000	0.875	0.500	2-0
0.6250	1.125	0.562	1
0.7500	1.250	0.625	2
-	±0.015	±0.015	-

Source: Ref. 3.

TABLE 7.6 Outside Diameter Tolerance

Diametral pitch	Tolerance, inch +0.000, -0.XXX
20	0.010
24	0.008
32	0.006
40	0.005
48	0.004
64	0.003
80	0.003
96	0.002
120	0.002

Module	Tolerance, mm +0.00, -0.XX
1.50	0.25
1.25	0.25
1.00	0.20
0.80	0.16
0.75	0.15
0.60	0.12
0.50	0.10
0.40	0.08
0.30	0.06
0.25	0.05

TABLE 7.7 Tooth Thickness Tolerance Classes for Spur, Helical, Herringbone, Rack, and Pinion (all tolerance values in inches)

Quality Number	Diametral Pitch	Class				
		A	B	C	D	E
Q3	.5	.074				
	1.2	.031				
	2.0	.019				
	3.2	.012				
	5.0	.0075				
Q4	.5	.074				
	1.2	.031				
	2.0	.019				
	3.2	.012				
	5.0	.0075				
Q5	.5	.074				
	1.2	.031				
	2.0	.019	.0093			
	3.2	.012	.006			
	5.0	.0075	.0037			
	8.0	.005	.0025			
Q6	.5	.074				
	1.2	.031				
	2.0	.019	.0093			
	3.2	.012	.006			
	5.0	.0075	.0037			
	8.0	.005	.0025			
	12.0	.003	.0018			
	20.0	.0024	.0012	.0006		
	32.0	.0016	.0008	.00043		
Q7 and higher	.5	.074				
	1.2	.031				
	2.0	.019	.0093	.0048		
	3.2	.012	.006	.003		
	5.0	.0075	.0037	.0019		
	8.0	.005	.0025	.00125	.00063	
	12.0	.003	.0018	.0009	.00044	
	20.0	.0024	.0012	.0006	.0003	.00016
	32.0	.0016	.0008	.00043	.0002	.0001
	50	.0012	.0006	.0003	.00014	.00007
	80	.0008	.00045	.00022	.00011	.000055
	120	.00067	.00034	.00017	.00009	.000045
	200	.0005	.00025	.00013	.00006	.00003

Source: Ref. 1.

TABLE 7.8 Coarse-Pitch Gear Tolerances (in ten-thousandths of an inch)

RUNOUT TOLERANCE

AGMA QUALITY NUMBER	NORMAL DIAMETRAL PITCH	PITCH DIAMETER (INCHES)									
		3/4	1½	3	6	12	25	50	100	200	400
Q3	1/2					788.2	938.6	1106.9	1305.5	1539.6	1815.7
	1				477.8	563.5	671.1	791.4	933.4	1100.8	1298.1
	2			289.7	341.6	402.9	479.8	565.9	667.4	787.1	928.2
	4			207.1	244.3	288.1	343.1	404.6	477.2	562.7	663.7
	8			148.1	174.7	206.0	245.3	289.3	341.2	402.4	474.5
Q4	1/2					563.0	670.4	790.7	932.5	1099.7	1297.0
	1				341.3	402.5	479.3	565.3	666.7	786.3	927.3
	2			206.9	244.0	287.8	342.7	404.2	476.7	562.2	663.0
	4			147.9	174.5	205.8	245.0	289.0	340.8	402.0	474.1
	8			105.8	124.8	147.1	175.2	206.6	243.7	287.4	338.9
Q5	1/2					402.1	478.9	564.8	666.1	785.5	926.4
	1				243.8	287.5	342.4	403.8	476.2	561.6	662.4
	2			147.8	174.3	205.6	244.8	288.7	340.5	401.6	473.6
	4		89.6	105.7	124.6	147.0	175.0	206.4	243.5	287.1	338.6
	8		64.1	75.6	89.1	105.1	125.1	147.6	174.1	205.3	242.1
Q6	1/2					287.2	342.1	403.4	475.8	561.1	661.7
	1				174.1	205.4	244.6	288.4	340.2	401.2	473.1
	2			105.6	124.5	146.8	174.9	206.2	243.2	286.8	338.3
	4		64.0	75.5	89.0	105.0	125.0	147.4	173.9	205.1	241.9
	8	38.8	45.8	54.0	63.6	75.1	89.4	105.4	124.3	146.6	172.9
	12	31.9	37.6	44.4	52.3	61.7	73.5	86.6	102.2	120.5	142.1
	20	24.9	29.4	34.6	40.8	48.2	57.4	67.7	79.8	94.1	111.0
Q7	1/2					205.2	244.3	288.1	339.8	400.8	472.7
	1				124.4	146.7	174.7	206.0	243.0	286.6	337.9
	2			75.4	88.9	104.9	124.9	147.3	173.7	204.9	241.6
	4		45.7	53.9	63.6	75.0	89.3	105.3	124.2	146.5	172.8
	8	27.7	32.7	38.5	45.5	53.6	63.9	75.3	88.8	104.7	123.5
	12	22.8	26.9	31.7	37.4	44.1	52.5	61.9	73.0	86.1	101.5
	20	17.8	21.0	24.7	29.2	34.4	41.0	48.3	57.0	67.2	79.3

PITCH TOLERANCE

AGMA QUALITY NUMBER	NORMAL DIAMETRAL PITCH	PITCH DIAMETER (INCHES)									
		3/4	1½	3	6	12	25	50	100	200	400
Q6	1/2					38.4	43.7	49.4	55.9	63.2	71.4
	1				29.1	32.9	37.4	42.3	47.8	54.1	61.1
	2			22.0	24.9	28.1	32.0	36.2	41.0	46.3	52.3
	4		16.7	18.9	21.3	24.1	27.4	31.0	35.1	39.6	44.8
	8	12.6	14.3	16.1	18.2	20.6	23.5	26.6	30.0	33.9	38.4
	12	11.5	13.0	14.7	16.7	18.8	21.5	24.3	27.4	31.0	35.0
	20	10.3	11.6	13.1	14.9	16.8	19.1	21.6	24.5	27.6	31.3
Q7	1/2					27.0	30.8	34.8	39.3	44.5	50.3
	1				20.5	23.1	26.4	29.8	33.7	38.1	43.1
	2			15.5	17.5	19.8	22.6	25.5	28.8	32.6	36.9
	4		11.7	13.3	15.0	17.0	19.3	21.8	24.7	29.9	31.6
	8	8.9	10.1	11.4	12.9	14.5	16.5	18.7	21.1	23.9	27.0
	12	8.1	9.2	10.4	11.7	13.3	15.1	17.1	19.3	21.8	24.7
	20	7.2	8.2	9.3	10.5	11.8	13.5	15.2	17.2	19.5	22.0

TABLE 7.8 (continued)

RUNOUT TOLERANCE — PITCH DIAMETER (INCHES)

AGMA Quality Number	Normal Diametral Pitch	3/4	1½	3	6	12	25	50	100	200	400
Q8	1/2					146.5	174.5	205.8	242.7	286.3	337.6
	1				88.8	104.8	124.8	147.2	173.6	204.7	241.4
	2			53.9	63.5	74.9	89.2	105.2	124.1	146.3	172.6
	4		32.7	38.5	45.4	53.6	63.8	75.2	88.7	104.6	123.4
	8	19.8	23.3	27.5	32.5	38.3	45.6	53.8	63.4	74.8	88.2
	12	16.3	19.2	22.6	26.7	31.5	37.5	44.2	52.1	61.5	72.5
	20	12.7	15.0	17.7	20.8	24.6	29.3	34.5	40.7	48.0	56.6
Q9	1/2					104.7	124.7	147.0	173.4	204.5	241.2
	1				63.5	74.8	89.1	105.1	124.0	146.2	172.4
	2			38.5	45.4	53.5	63.7	75.2	88.6	104.5	123.3
	4		23.3	27.5	32.4	38.3	45.6	53.7	63.4	74.7	88.1
	8	14.1	16.7	19.7	23.2	27.4	32.6	38.4	45.3	53.4	63.0
	12	11.6	13.7	16.2	19.1	22.5	26.8	31.6	37.2	43.9	51.8
	20	9.1	10.7	12.6	14.9	17.6	20.9	24.7	29.1	34.3	40.4
Q10	1/2					75.1	89.0	105.0	123.8	146.1	172.3
	1				45.3	53.5	63.7	75.1	88.5	104.4	123.0
	2			27.5	32.5	38.2	45.5	53.7	63.3	74.7	88.1
	4		16.7	19.6	23.2	27.5	32.5	38.4	45.3	53.4	63.0
	8	10.1	11.9	14.0	16.6	19.5	23.3	27.4	32.4	38.2	45.0
	12	8.3	9.8	11.5	13.6	16.1	19.1	22.6	26.6	31.4	37.0
	20	6.5	7.6	9.0	10.6	12.5	14.9	17.6	20.8	24.7	28.9
Q11	1/2					53.6	63.6	75.0	88.0	104.3	123.0
	1				32.4	38.2	45.5	53.6	63.2	74.6	87.9
	2			19.6	23.2	27.3	32.5	38.4	45.2	53.4	62.9
	4		11.9	14.0	16.6	19.6	23.2	27.4	32.4	38.1	45.0
	8	7.2	8.5	10.0	11.9	13.9	16.6	19.6	23.1	27.3	32.1
	12	5.9	7.0	8.2	9.7	11.5	13.6	16.1	19.0	22.4	26.4
	20	4.6	5.4	6.4	7.6	8.9	10.6	12.6	14.9	17.6	20.6
Q12	1/2					38.3	45.4	53.6	62.9	74.5	87.9
	1				23.1	27.3	32.5	38.3	45.2	53.3	62.8
	2			14.0	16.6	19.5	23.2	27.4	32.3	38.1	44.9
	4		8.5	10.0	11.9	14.0	16.5	19.5	23.1	27.2	32.1
	8	5.2	6.1	7.2	8.5	9.9	11.9	14.0	16.5	19.5	23.0
	12	4.2	5.0	5.9	6.9	8.2	9.7	11.5	13.6	16.0	18.9
	20	3.3	3.9	4.6	5.4	6.4	7.6	8.9	10.6	12.5	14.7

PITCH TOLERANCE — PITCH DIAMETER (INCHES)

AGMA Quality Number	Normal Diametral Pitch	3/4	1½	3	6	12	25	50	100	200	400
Q8	1/2					19.0	21.7	24.5	27.7	31.3	35.4
	1				14.4	16.3	18.6	21.0	23.7	26.8	30.3
	2			10.9	12.3	14.0	15.9	18.0	20.3	23.0	26.0
	4		8.3	9.3	10.6	11.9	13.6	15.4	17.4	19.7	22.2
	8	7.1	8.0	9.0	10.2	11.7	13.2	14.9	16.8	19.0	
	12	6.5	7.3	8.3	9.3	10.6	12.0	13.6	15.4	17.4	
	20	5.8	6.5	7.4	8.3	9.5	10.7	12.1	13.7	15.5	
Q9	1/2					13.4	15.3	17.3	19.5	22.1	24.9
	1				10.2	11.5	13.1	14.8	16.7	18.9	21.4
	2			7.7	8.7	9.8	11.2	12.7	14.3	16.2	18.3
	4		5.8	6.6	7.4	8.4	9.6	10.8	12.2	13.8	15.7
	8	5.0	5.6	6.4	7.2	8.2	9.3	10.5	11.9	13.4	
	12	4.6	5.1	5.8	6.6	7.5	8.5	9.6	10.8	12.2	
	20	4.1	4.6	5.2	5.9	6.7	7.6	8.5	9.7	10.9	
Q10	1/2					9.6	10.8	12.2	13.8	15.7	17.6
	1				7.2	8.2	9.3	10.5	11.9	13.4	15.1
	2			5.4	6.1	6.9	7.9	8.9	10.1	11.4	12.8
	4		4.1	4.6	5.2	5.9	6.7	7.6	8.5	9.7	11.0
	8	3.5	3.9	4.5	5.1	5.8	6.5	7.3	8.3	9.4	
	12	3.2	3.6	4.1	4.6	5.3	6.0	6.7	7.6	8.6	
	20	2.9	3.3	3.7	4.2	4.7	5.4	6.0	6.8	7.6	
Q11	1/2					6.9	7.9	8.9	10.1	11.4	12.9
	1				5.1	5.9	6.6	7.5	8.5	9.6	10.8
	2			3.9	4.4	4.9	5.6	6.4	7.2	8.1	9.1
	4		2.9	3.3	3.7	4.2	4.8	5.4	6.1	6.9	7.9
	8	2.5	2.8	3.2	3.6	4.1	4.6	5.3	5.9	6.7	
	12	2.3	2.6	2.9	3.3	3.8	4.3	4.8	5.4	6.1	
	20	2.0	2.3	2.6	2.9	3.3	3.7	4.3	4.8	5.4	
Q12	1/2					4.9	5.5	6.2	7.1	8.0	9.0
	1				3.6	4.2	4.7	5.4	6.1	6.9	7.7
	2			2.8	3.1	3.5	4.0	4.6	5.1	5.8	6.5
	4		2.0	2.3	2.6	3.0	3.3	3.8	4.3	4.8	5.4
	8	1.8	2.0	2.3	2.6	2.9	3.3	3.8	4.1	4.8	
	12	1.7	1.9	2.0	2.3	2.6	3.0	3.3	3.8	4.3	
	20	1.4	1.6	1.8	2.0	2.3	2.6	3.0	3.4	3.9	

PROFILE TOLERANCE — PITCH DIAMETER (INCHES)

AGMA Quality Number	Normal Diametral Pitch	1½	3	6	12	25	50	100	200	400
Q8	1/2				42.6	47.7	53.1	59.1	65.7	73.1
	1			28.3	31.5	35.3	39.3	43.7	48.6	54.1
	2		18.8	21.0	23.3	26.1	29.0	32.3	36.0	40.0
	4	12.5	13.9	15.5	17.2	19.3	21.5	23.9	26.6	29.6
	8	9.3	10.3	11.5	12.8	14.3	15.9	17.7	19.7	21.9
	12	7.8	8.6	9.6	10.7	12.0	13.3	14.8	16.5	18.4
	20	6.2	6.9	7.7	8.6	9.6	10.7	11.9	13.2	14.7
Q9	1/2				30.4	34.1	37.9	42.2	46.9	52.2
	1			20.2	22.5	25.2	28.1	31.2	34.7	38.6
	2		13.5	15.0	16.7	18.6	20.7	23.1	25.7	28.6
	4	8.9	10.0	11.1	12.3	13.8	15.3	17.1	19.0	21.1
	8	6.6	7.4	8.2	9.1	10.2	11.4	12.6	14.1	15.6
	12	5.5	6.2	6.9	7.6	8.6	9.5	10.6	11.8	13.1
	20	4.4	4.9	5.5	6.1	6.8	7.6	8.5	9.4	10.5
Q10	1/2				21.7	24.3	27.1	30.1	33.5	37.3
	1			14.5	16.1	18.0	20.0	22.3	24.8	27.6
	2		9.6	10.7	11.9	13.3	14.8	16.5	18.3	20.4
	4	6.4	7.1	7.9	8.8	9.9	11.0	12.2	13.6	15.1
	8	4.7	5.3	5.9	6.5	7.3	8.1	9.0	10.0	11.2
	12	4.0	4.4	4.9	5.5	6.1	6.8	7.6	8.4	9.4
	20	3.2	3.5	3.9	4.4	4.9	5.4	6.1	6.7	7.5
Q11	1/2				15.5	17.4	19.3	21.5	24.0	26.7
	1			10.3	11.5	12.9	14.3	15.9	17.7	19.7
	2		6.9	7.6	8.5	9.5	10.6	11.8	13.1	14.6
	4	4.6	5.1	5.6	6.3	7.0	7.8	8.7	9.7	10.8
	8	3.4	3.8	4.2	4.6	5.2	5.8	6.4	7.2	8.0
	12	2.8	3.1	3.5	3.9	4.4	4.9	5.4	6.0	6.7
	20	2.3	2.5	2.8	3.1	3.5	3.9	4.3	4.8	5.4
Q12	1/2				11.1	12.4	13.8	15.4	17.1	19.0
	1			7.4	8.2	9.2	10.2	11.4	12.7	14.1
	2		4.9	5.5	6.1	6.8	7.6	8.4	9.4	10.4
	4	3.3	3.6	4.0	4.5	5.0	5.6	6.2	6.9	7.7
	8	2.4	2.7	3.0	3.3	3.7	4.1	4.6	5.1	5.7
	12	2.0	2.2	2.5	2.7	3.1	3.5	3.9	4.3	4.8
	20	1.8	2.0	2.2	2.5	2.8	3.1	3.4	3.8	3.8

LEAD TOLERANCE — FACE WIDTH (INCHES)

AGMA Quality Number	1 and Less	2	3	4	5
Q8	5	8	11	13	16
Q9	4	7	9	11	13
Q10	3	5	7	9	10
Q11	3	4	6	7	8
Q12	2	3	5	6	7

(continued)

TABLE 7.8 (continued)

AGMA QUALITY NUMBER	NORMAL DIAMETRAL PITCH	RUNOUT TOLERANCE — PITCH DIAMETER (INCHES)										PITCH TOLERANCE — PITCH DIAMETER (INCHES)										PROFILE TOLERANCE — PITCH DIAMETER (INCHES)										LEAD TOLERANCE — FACE WIDTH (INCHES)				
		3/4	1½	3	6	12	25	50	100	200	400	3/4	1½	3	6	12	25	50	100	200	400	3/4	1½	3	6	12	25	50	100	200	400	1 and Less	2	3	4	5
Q13	1/2					27.2	32.4	38.3	45.1	53.2	62.8					3.3	3.8	4.2	4.8	5.4	6.1					7.9	8.9	9.9	11.0	12.2	13.6	2	3	4	4	5
	1				16.5	19.5	23.2	27.4	32.3	38.1	44.9				2.5	2.8	3.2	3.6	4.1	4.6	5.3				5.3	5.9	6.6	7.3	8.1	9.0	10.1					
	2			10.0	11.8	13.9	16.6	19.6	23.1	27.2	32.1			1.9	2.1	2.4	2.8	3.1	3.5	4.0	4.5			3.5	3.9	4.3	4.9	5.4	6.0	6.7	7.4					
	4		6.1	7.2	8.4	10.0	11.9	14.0	16.5	19.5	22.9		1.4	1.6	1.8	2.1	2.4	2.7	3.0	3.4	3.8		2.3	2.6	2.9	3.2	3.6	4.0	4.4	4.9	5.5					
	8	3.7	4.3	5.1	6.0	7.1	8.5	10.0	11.8	13.9	16.4	1.1	1.3	1.4	1.6	1.8	2.0	2.3	2.6	2.9	3.3	1.5	1.7	1.9	2.1	2.4	2.7	3.0	3.3	3.7	4.1					
	12	3.0	3.6	4.2	5.0	5.9	7.0	8.2	9.7	11.4	13.5	1.0	1.1	1.3	1.4	1.6	1.8	2.1	2.4	2.7	3.0	1.3	1.4	1.6	1.8	2.0	2.2	2.5	2.8	3.1	3.4					
	20	2.4	2.8	3.3	3.9	4.6	5.4	6.4	7.6	8.9	10.5	0.9	1.0	1.1	1.3	1.4	1.6	1.9	2.1	2.4	2.7	1.0	1.2	1.3	1.4	1.6	1.8	2.0	2.2	2.5	2.7					
Q14	1/2					19.5	23.2	27.3	32.2	38.0	44.8					2.3	2.6	3.0	3.4	3.8	4.3					5.7	6.3	7.1	7.8	8.7	9.7	1	2	3	4	4
	1				11.8	13.9	16.6	19.5	23.0	27.2	32.1				1.8	2.0	2.3	2.6	2.9	3.3	3.7				3.8	4.2	4.7	5.2	5.8	6.5	7.2					
	2			7.2	8.4	9.9	11.8	14.0	16.5	19.4	22.9			1.3	1.5	1.7	1.9	2.2	2.5	2.8	3.2			2.5	2.8	3.1	3.5	3.9	4.3	4.8	5.3					
	4		4.3	5.1	6.0	7.1	8.5	10.0	11.8	13.9	16.4	1.0	1.0	1.1	1.3	1.5	1.7	1.9	2.1	2.4	2.7	1.7	1.7	1.9	2.1	2.3	2.6	2.9	3.2	3.5	3.9					
	8	2.6	3.1	3.7	4.3	5.1	6.1	7.1	8.4	9.9	11.7	0.8	0.9	1.0	1.1	1.2	1.4	1.6	1.8	2.1	2.3	1.1	1.2	1.5	1.5	1.7	1.9	2.1	2.3	2.6	2.9					
	12	2.2	2.5	3.0	3.5	4.2	5.0	5.9	6.9	8.2	9.6	0.7	0.8	0.9	1.0	1.0	1.3	1.5	1.7	1.9	2.1	0.9	1.0	1.3	1.3	1.4	1.6	1.8	2.0	2.2	2.4					
	20	1.7	2.0	2.3	2.8	3.3	3.9	4.6	5.4	6.4	7.5	0.6	0.7	0.8	0.9	1.0	1.1	1.3	1.5	1.7	1.9	0.7	0.8	0.9	1.0	1.0	1.1	1.4	1.4	1.6	1.8					
Q15	1/2					13.9	16.6	19.5	23.0	27.2	32.0					1.6	1.9	2.1	2.4	2.7	3.0					4.0	4.5	5.0	5.6	6.2	6.9	1	2	2	3	3
	1				8.4	9.9	11.8	14.0	16.5	19.4	22.9				1.2	1.4	1.6	1.8	2.0	2.3	2.6				2.7	3.0	3.3	3.7	4.1	4.6	5.1					
	2			5.1	6.0	7.1	8.5	10.0	11.8	13.9	16.4			0.9	1.1	1.2	1.4	1.5	1.7	2.0	2.2			1.8	2.0	2.2	2.5	2.8	3.1	3.4	3.8					
	4		3.1	3.7	4.3	5.1	6.1	7.1	8.4	9.9	11.7	0.7	0.8	0.9	1.0	1.0	1.2	1.3	1.5	1.7	1.9	1.2	1.3	1.5	1.6	1.8	2.0	2.3	2.5	2.8					
	8	1.9	2.2	2.6	3.1	3.6	4.3	5.1	6.0	7.1	8.4	0.6	0.6	0.7	0.8	0.9	1.0	1.1	1.3	1.4	1.6	0.9	0.9	1.0	1.1	1.2	1.4	1.5	1.7	1.9	2.1					
	12	1.5	1.8	3.0	2.5	3.0	3.6	4.2	4.9	5.8	6.9	0.6	0.6	0.6	0.7	0.8	0.9	1.0	1.2	1.3	1.5	0.7	0.7	0.9	0.9	1.0	1.1	1.3	1.4	1.6	1.7					
	20	1.2	1.4	1.7	2.0	2.3	2.8	3.3	3.9	4.6	5.4	0.5	0.5	0.6	0.6	0.7	0.8	0.9	1.0	1.2	1.3	0.5	0.6	0.7	0.7	0.8	0.9	1.0	1.1	1.3	1.4					

Source: Ref. 1.

TABLE 7.9 Total Composite Tolerance (tolerances in ten-thousandths of an inch)

AGMA Quality Number	Normal Diametral Pitch	0.040	0.063	0.100	0.160	0.250	0.400	0.630	1.0	1.6	2.5	4.0	6.3	10.	16.	25.	50.	100.	200.
Q5	0.5																		
	1.0																		
	2.0																		
	4.0																		
	8.0																		
	12.0							94.8	95.0	95.4	98.4	106.4	114.7	123.8	133.8	144.0	161.5	181.2	
	20.0						74.3	74.3	74.3	77.0	82.5	88.6	95.0	102.0	109.6	117.3	130.5		
	24.0					68.3	68.3	68.2	67.6	72.5	77.4	83.1	88.9	95.2	102.1	109.1	121.0		
	30.0					61.7	61.6	61.4	62.8	67.2	71.7	76.7	81.9	87.6	93.7	99.9			
	32.0					60.0	59.8	59.6	61.5	65.8	70.1	75.0	80.0	85.5	91.4	97.4			
	40.0				54.4	54.3	54.0	53.7	57.2	61.1	65.0	69.3	73.8	78.7	83.9	89.3			
	48.0				50.2	50.0	49.7	50.7	54.0	57.5	61.0	65.0	69.1	73.5	78.3	83.1			
	50.0				49.3	49.1	48.8	50.1	53.2	56.7	60.2	64.1	68.1	72.4	77.1				
	60.0			45.8	45.6	45.3	44.6	47.3	50.2	53.3	56.5	60.1	63.7	67.7	71.9				
	64.0			44.5	44.3	44.0	43.7	46.3	49.1	52.2	55.3	58.7	62.3	66.1	70.2				
	80.0			40.5	40.3	39.9	40.8	43.2	45.7	48.5	51.2	54.3	57.5	60.9					
	100.0		37.1	36.9	36.5	36.0	38.1	40.2	42.5	45.0	47.5	50.2	53.1	56.1					
	200.0	28.2	27.9	27.4	27.8	29.2	30.7	32.3	34.0	35.7	37.5	39.5							

(continued)

TABLE 7. 9 (continued)

AGMA Quality Number	Normal Diametral Pitch	0.040	0.063	0.100	0.160	0.250	0.400	0.630	1.0	1.6	2.5	4.0	6.3	10.	16.	25.	50.	100.	200.
Q6	0.5														434.3	447.5	480.3	577.7	694.8
	1.0												268.5	273.8	282.3	293.0	346.1	408.7	482.7
	2.0											176.3	179.4	184.4	196.9	216.8	251.9	292.6	340.0
	4.0									118.4	119.4	121.2	123.6	135.2	148.2	161.7	185.0	211.8	242.5
	8.0								82.6	83.1	83.9	87.9	95.2	103.3	112.2	121.3	137.1	154.8	
	12.0							67.5	67.7	68.1	70.3	76.0	81.9	88.4	95.6	102.9	115.4	129.4	
	20.0						53.0	53.0	53.0	55.0	58.9	63.3	67.9	72.8	78.3	83.8	93.2		
	24.0					48.8	48.8	48.7	48.3	51.8	55.3	59.3	63.5	68.0	72.9	78.0	86.4		
	30.0					44.2	44.1	43.9	44.9	48.0	51.2	54.8	58.5	62.6	66.9	71.4			
	32.0					42.9	42.8	42.6	44.0	47.0	50.1	53.6	57.2	61.1	65.3	69.6			
	40.0				39.0	38.9	38.6	38.4	40.9	43.6	46.4	49.5	52.7	56.2	59.9	63.8			
	48.0				36.0	35.8	35.5	36.2	38.5	41.1	43.6	46.4	49.3	52.5	55.9	59.4			
	50.0				35.4	35.2	34.9	35.8	38.0	40.5	43.0	45.8	48.6	51.7	55.1				
	60.0			32.9	32.7	32.4	31.8	33.8	35.8	38.1	40.4	42.9	45.5	48.3	51.4				
	64.0			32.0	31.8	31.5	31.2	33.1	35.1	37.3	39.5	42.0	44.5	47.2	50.1				
	80.0			29.1	28.9	28.5	29.1	30.8	32.7	34.6	36.6	38.8	41.1	43.5					
	100.0		26.7	26.5	26.2	25.7	27.2	28.7	30.4	32.1	33.9	35.9	37.9	40.1					
	200.0	20.4	20.1	19.6	19.9	20.9	22.0	23.1	24.3	25.5	26.8	28.2							

AGMA Quality Number	Normal Diametral Pitch	0.040	0.063	0.100	0.160	0.250	0.400	0.630	1.0	1.6	2.5	4.0	6.3	10.	16.	25.	50.	100.	200.
Q7	0.5														300.2	313.4	343.1	412.6	496.3
	1.0												186.2	191.5	200.0	209.3	247.2	291.9	344.8
	2.0											123.6	126.7	131.7	140.6	154.9	179.9	209.0	242.8
	4.0									83.4	84.5	86.3	88.3	96.6	105.8	115.5	132.2	151.3	173.2
	8.0								58.6	59.1	59.9	62.8	68.0	73.8	80.1	86.7	97.9	110.6	
	12.0							48.0	48.3	48.6	50.2	54.3	58.5	63.2	68.3	73.5	82.4	92.4	
	20.0						37.9	37.9	37.9	39.3	42.1	45.2	48.5	52.0	55.9	59.9	66.6		
	24.0					34.9	34.9	34.8	34.5	37.0	39.5	42.4	45.3	48.6	52.1	55.7	61.7		
	30.0					31.6	31.5	31.4	32.1	34.3	36.6	39.2	41.8	44.7	47.8	51.0			
	32.0					30.8	30.6	30.4	31.4	33.6	35.8	38.3	40.8	43.6	46.6	49.7			
	40.0				28.0	27.9	27.6	27.4	29.2	31.2	33.1	35.4	37.7	40.1	42.8	45.5			
	48.0				25.9	25.7	25.4	25.9	27.5	29.3	31.1	33.2	35.2	37.5	39.9	42.4			
	50.0				25.4	25.2	24.9	25.5	27.2	28.9	30.7	32.7	34.7	36.9	39.3				
	60.0			23.7	23.5	23.2	22.7	24.1	25.6	27.2	28.8	30.7	32.5	34.5	36.7				
	64.0			23.1	22.9	22.6	22.3	23.6	25.1	26.6	28.2	30.0	31.8	33.7	35.8				
	80.0			21.0	20.7	20.3	20.8	22.0	23.3	24.7	26.1	27.7	29.3	31.1					
	100.0		19.3	19.1	18.8	18.4	19.4	20.5	21.7	23.0	24.2	25.6	27.1	28.6					
	200.0	14.8	14.5	14.0	14.2	14.9	15.7	16.5	17.3	18.2	19.1	20.1							

(continued)

189

TABLE 7. 9 (continued)

AGMA Quality Number	Normal Diametral Pitch	0.040	0.063	0.100	0.160	0.250	0.400	0.630	1.0	1.6	2.5	4.0	6.3	10.	16.	25.	50.	100.	200.
Q8	0.5														204.4	217.6	245.1	294.7	354.5
	1.0												127.4	132.7	141.2	149.5	176.6	208.5	246.3
	2.0											86.0	89.1	94.1	100.4	110.6	128.5	149.3	173.5
	4.0										59.5	61.3	63.0	69.0	75.6	82.5	94.4	108.1	123.7
	8.0								41.5	42.0	42.8	44.9	48.6	52.7	57.2	61.9	69.9	79.0	
	12.0							34.1	34.4	34.7	35.9	38.8	41.8	45.1	48.8	52.5	58.9	66.0	
	20.0						27.1	27.1	27.1	28.1	30.1	32.3	34.6	37.2	39.9	42.8	47.6		
	24.0					25.0	25.0	24.9	24.6	26.4	28.2	30.3	32.4	34.7	37.2	39.8	44.1		
	30.0					22.7	22.6	22.4	22.9	24.5	26.1	28.0	29.9	31.9	34.2	36.4			
	32.0					22.1	21.9	21.7	22.4	24.0	25.6	27.3	29.2	31.2	33.3	35.5			
	40.0				20.1	20.0	19.8	19.6	20.9	22.3	23.7	25.3	26.9	28.7	30.6	32.5			
	48.0				18.6	18.5	18.1	18.5	19.7	20.9	22.2	23.7	25.2	26.8	28.5	30.3			
	50.0				18.3	18.1	17.8	18.2	19.4	20.7	21.9	23.3	24.8	26.4	28.1				
	60.0			17.1	16.9	16.7	16.2	17.2	18.3	19.4	20.6	21.9	23.2	24.7	26.2				
	64.0			16.7	16.5	16.2	15.9	16.9	17.9	19.0	20.2	21.4	22.7	24.1	25.6				
	80.0			15.2	14.9	14.5	14.9	15.7	16.7	17.7	18.7	19.8	20.9	22.2					
	100.0		14.0	13.8	13.5	13.1	13.9	14.7	15.5	16.4	17.3	18.3	19.3	20.4					
	200.0	10.8	10.5	10.0	10.1	10.6	11.2	11.8	12.4	13.0	13.7	14.4							

AGMA Quality Number	Normal Diametral Pitch	0.040	0.063	0.100	0.160	0.250	0.400	0.630	1.0	1.6	2.5	4.0	6.3	10.	16.	25.	50.	100.	200.
Q9	0.5														136.0	149.1	175.0	210.5	253.2
	1.0												85.4	90.7	99.3	106.8	126.1	148.9	175.9
	2.0											59.1	62.2	67.2	71.7	79.0	91.8	106.7	123.9
	4.0									40.5	41.6	43.4	45.0	49.3	54.0	58.9	67.4	77.2	88.4
	8.0								29.2	29.8	30.6	32.0	34.7	37.6	40.9	44.2	49.9	56.4	
	12.0							24.2	24.4	24.8	25.6	27.7	29.9	32.2	34.8	37.5	42.0	47.2	
	20.0						19.3	19.3	19.2	20.1	21.5	23.1	24.7	26.5	28.5	30.5	34.0		
	24.0					17.9	17.9	17.8	17.6	18.9	20.2	21.6	23.1	24.8	26.6	28.4	31.5		
	30.0					16.3	16.2	16.0	16.4	17.5	18.7	20.0	21.3	22.8	24.4	26.0			
	32.0					15.9	15.7	15.5	16.0	17.1	18.3	19.5	20.8	22.3	23.8	25.4			
	40.0				14.5	14.4	14.2	14.0	14.9	15.9	16.9	18.0	19.2	20.5	21.8	23.2			
	48.0				13.5	13.3	13.0	13.2	14.0	15.0	15.9	16.9	18.0	19.1	20.4	21.6			
	50.0				13.2	13.0	12.7	13.0	13.9	14.8	15.7	16.7	17.7	18.8	20.1				
	60.0			12.4	12.3	12.0	11.6	12.3	13.1	13.9	14.7	15.6	16.6	17.6	18.7				
	64.0			12.1	11.9	11.6	11.4	12.1	12.8	13.6	14.4	15.3	16.2	17.2	18.3				
	80.0			11.0	10.8	10.4	10.6	11.2	11.9	12.6	13.3	14.1	15.0	15.8					
	100.0		10.3	10.0	9.7	9.4	9.9	10.5	11.1	11.7	12.4	13.1	13.8	14.6					
	200.0	7.9	7.6	7.1	7.2	7.6	8.0	8.4	8.8	9.3	9.8	10.3							

(continued)

191

TABLE 7. 9 (continued)

AGMA Quality Number	Normal Diametral Pitch	0.040	0.063	0.100	0.160	0.250	0.400	0.630	1.0	1.6	2.5	4.0	6.3	10.	16.	25.	50.	100.	200.
Q10	0.5														87.1	100.3	125.0	150.4	180.9
	1.0												55.5	60.7	69.3	76.3	90.1	106.4	125.6
	2.0											39.9	43.0	48.0	51.2	56.4	65.6	76.2	88.5
	4.0									27.8	28.9	30.7	32.2	35.2	38.6	42.1	48.2	55.1	63.1
	8.0								20.5	21.0	21.8	22.9	24.8	26.9	29.2	31.6	35.7	40.3	
	12.0							17.1	17.3	17.7	18.3	19.8	21.3	23.0	24.9	26.8	30.0	33.7	
	20.0						13.8	13.8	13.8	14.3	15.3	16.5	17.7	19.0	20.4	21.8	24.3		
	24.0					12.8	12.8	12.7	12.6	13.5	14.4	15.4	16.5	17.7	19.0	20.3	22.5		
	30.0					11.7	11.6	11.4	11.7	12.5	13.3	14.3	15.2	16.3	17.4	18.6			
	32.0					11.4	11.3	11.1	11.4	12.2	13.0	13.9	14.9	15.9	17.0	18.1			
	40.0				10.5	10.4	10.2	10.0	10.6	11.4	12.1	12.9	13.7	14.6	15.6	16.6			
	48.0				9.8	9.6	9.3	9.4	10.0	10.7	11.3	12.1	12.8	13.7	14.6	15.5			
	50.0				9.6	9.4	9.1	9.3	9.9	10.5	11.2	11.9	12.7	13.5	14.3				
	60.0			9.1	8.9	8.6	8.3	8.8	9.3	9.9	10.5	11.2	11.9	12.6	13.4				
	64.0			8.9	8.7	8.4	8.1	8.6	9.1	9.7	10.3	10.9	11.6	12.3	13.1				
	80.0			8.1	7.8	7.4	7.6	8.0	8.5	9.0	9.5	10.1	10.7	11.3					
	100.0		7.6	7.3	7.0	6.7	7.1	7.5	7.9	8.4	8.8	9.3	9.9	10.4					
	200.0	5.9	5.6	5.1	5.2	5.4	5.7	6.0	6.3	6.6	7.0	7.3							

(continued)

AGMA Quality Number	Normal Diametral Pitch	0.040	0.063	0.100	0.160	0.250	0.400	0.630	1.0	1.6	2.5	4.0	6.3	10.	16.	25.	50.	100.	200.
Q11	0.5														52.2	65.3	89.3	107.4	129.2
	1.0												34.0	39.3	47.9	54.5	64.3	76.3	89.7
	2.0											26.2	29.3	34.3	36.6	40.3	46.8	54.4	63.2
	4.0									18.7	19.8	21.6	23.0	25.1	27.6	30.1	34.4	39.4	45.1
	8.0								14.3	14.8	15.6	16.4	17.7	19.2	20.9	22.6	25.5	28.8	
	12.0							12.0	12.3	12.6	13.1	14.1	15.2	16.4	17.8	19.1	21.5	24.1	
	20.0						9.9	9.9	9.9	10.2	11.0	11.8	12.6	13.5	14.6	15.6	17.3		
	24.0					9.2	9.2	9.1	9.0	9.6	10.3	11.0	11.8	12.6	13.6	14.5	16.1		
	30.0					8.5	8.4	8.2	8.3	8.9	9.5	10.2	10.9	11.6	12.4	13.3			
	32.0					8.3	8.1	7.9	8.2	8.7	9.3	10.0	10.6	11.4	12.1	12.9			
	40.0				7.7	7.5	7.3	7.1	7.6	8.1	8.6	9.2	9.8	10.4	11.1	11.9			
	48.0				7.1	6.9	6.6	6.7	7.2	7.6	8.1	8.6	9.2	9.8	10.4	11.0			
	50.0				7.0	6.8	6.5	6.6	7.1	7.5	8.0	8.5	9.0	9.6	10.2				
	60.0			6.7	6.5	6.2	5.9	6.3	6.7	7.1	7.5	8.0	8.5	9.0	9.6				
	64.0			6.5	6.3	6.0	5.8	6.2	6.5	6.9	7.3	7.8	8.3	8.8	9.3				
	80.0			6.0	5.7	5.3	5.4	5.7	6.1	6.4	6.8	7.2	7.6	8.1					
	100.0		5.6	5.4	5.1	4.8	5.1	5.3	5.6	6.0	6.3	6.7	7.0	7.4					
	200.0	4.4	4.1	3.6	3.7	3.9	4.1	4.3	4.5	4.7	5.0	5.2							

193

TABLE 7.9 (continued)

AGMA Quality Number	Normal Diametral Pitch	0.040	0.063	0.100	0.160	0.250	0.400	0.630	1.0	1.6	2.5	4.0	6.3	10.	16.	25.	50.	100.	200.
	0.5														27.2	40.4	63.8	76.7	92.3
	1.0												18.7	24.0	32.6	38.9	46.0	54.3	64.1
	2.0											16.4	19.5	24.5	26.1	28.8	33.5	38.9	45.2
	4.0									12.2	13.3	15.1	16.4	18.0	19.7	21.5	24.6	28.1	32.2
	8.0								9.8	10.3	11.1	11.7	12.6	13.7	14.9	16.1	18.2	20.6	
	12.0							8.4	8.6	9.0	9.3	10.1	10.9	11.7	12.7	13.7	15.3	17.2	
	20.0						7.0	7.0	7.0	7.3	7.8	8.4	9.0	9.7	10.4	11.1	12.4		
	24.0					6.6	6.6	6.5	6.4	6.9	7.3	7.9	8.4	9.0	9.7	10.4	11.5		
Q12	30.0					6.1	6.0	5.9	6.0	6.4	6.8	7.3	7.8	8.3	8.9	9.5			
	32.0					6.0	5.9	5.7	5.8	6.2	6.7	7.1	7.6	8.1	8.7	9.2			
	40.0				5.6	5.5	5.3	5.1	5.4	5.8	6.2	6.6	7.0	7.5	8.0	8.5			
	48.0				5.3	5.1	4.7	4.8	5.1	5.5	5.8	6.2	6.6	7.0	7.4	7.9			
	50.0				5.2	5.0	4.6	4.7	5.1	5.4	5.7	6.1	6.5	6.9	7.3				
	60.0			5.0	4.8	4.5	4.2	4.5	4.8	5.1	5.4	5.7	6.0	6.4	6.8				
	64.0			4.9	4.7	4.4	4.1	4.4	4.7	5.0	5.2	5.6	5.9	6.3	6.7				
	80.0			4.5	4.2	3.8	3.9	4.1	4.3	4.6	4.9	5.2	5.5	5.8					
	100.0		4.3	4.0	3.7	3.4	3.6	3.8	4.0	4.3	4.5	4.8	5.0	5.3					
	200.0	3.4	3.1	2.6	2.6	2.8	2.9	3.1	3.2	3.4	3.6	3.7							

AGMA Quality Number	Normal Diametral Pitch	0.040	0.063	0.100	0.160	0.250	0.400	0.630	1.0	1.6	2.5	4.0	6.3	10.	16.	25.	50.	100.	200.
	0.5														9.4	22.6	45.6	54.8	65.9
	1.0												7.8	13.1	21.6	27.8	32.8	38.8	45.8
	2.0											9.4	12.5	17.5	18.7	20.6	23.9	27.8	32.3
	4.0									7.5	8.6	10.4	11.7	12.8	14.1	15.3	17.6	20.1	23.0
	8.0								6.6	7.1	8.0	8.3	9.0	9.8	10.6	11.5	13.0	14.7	
	12.0							5.8	6.1	6.4	6.7	7.2	7.8	8.4	9.1	9.8	10.9	12.3	
Q13	20.0						5.0	5.0	5.0	5.2	5.6	6.0	6.4	6.9	7.4	8.0	8.8		
	24.0					4.8	4.7	4.7	4.6	4.9	5.2	5.6	6.0	6.5	6.9	7.4	8.2		
	30.0					4.5	4.4	4.2	4.3	4.6	4.9	5.2	5.6	5.9	6.3	6.8			
	32.0					4.4	4.2	4.0	4.2	4.5	4.8	5.1	5.4	5.8	6.2	6.6			
	40.0				4.2	4.0	3.8	3.6	3.9	4.1	4.4	4.7	5.0	5.3	5.7	6.0			
	48.0				3.9	3.7	3.4	3.4	3.7	3.9	4.1	4.4	4.7	5.0	5.3	5.6			
	50.0				3.8	3.6	3.3	3.4	3.6	3.8	4.1	4.3	4.6	4.9	5.2				
	60.0			3.8	3.6	3.3	3.0	3.2	3.4	3.6	3.8	4.1	4.3	4.6	4.9				
	64.0			3.7	3.5	3.2	3.0	3.1	3.3	3.5	3.7	4.0	4.2	4.5	4.8				
	80.0			3.4	3.1	2.7	2.8	2.9	3.1	3.3	3.5	3.7	3.9	4.1					
	100.0		3.3	3.1	2.7	2.4	2.6	2.7	2.9	3.0	3.2	3.4	3.6	3.8					
	200.0	2.7	2.4	1.9	1.9	2.0	2.1	2.2	2.3	2.4	2.5	2.7							

(continued)

195

TABLE 7.9 (continued)

AGMA Quality Number	Normal Diametral Pitch	0.040	0.063	0.100	0.160	0.250	0.400	0.630	1.0	1.6	2.5	4.0	6.3	10.	16.	25.	50.	100.	200.
Q14	0.5															9.9	32.5	39.1	47.1
	1.0														13.8	19.9	23.5	27.7	32.7
	2.0											4.4	7.5	5.3	13.3	14.7	17.1	19.8	23.0
	4.0										5.3	7.1	8.4	12.5	10.0	11.0	12.5	14.4	16.4
	8.0								4.3	4.2	5.7	6.0	6.5	9.2	7.6	8.2	9.3	10.5	
	12.0							4.0	4.2	4.6	4.8	5.1	5.6	7.0	6.5	7.0	7.8	8.8	
	20.0						3.6	3.6	3.6	3.7	4.0	4.3	4.6	6.0	5.3	5.7	6.3		
	24.0					3.5	3.4	3.4	3.3	3.5	3.7	4.0	4.3	4.9	4.9	5.3	5.9		
	30.0					3.3	3.2	3.0	3.0	3.3	3.5	3.7	4.0	4.6	4.5	4.8			
	32.0					3.2	3.1	2.9	3.0	3.2	3.4	3.6	3.9	4.2	4.4	4.7			
	40.0				3.1	3.0	2.8	2.6	2.8	3.0	3.1	3.4	3.6	4.1	4.1	4.3			
	48.0				2.9	2.8	2.4	2.5	2.6	2.8	3.0	3.1	3.3	3.8	3.8	4.0			
	50.0				2.9	2.7	2.4	2.4	2.6	2.7	2.9	3.1	3.3	3.6	3.7				
	60.0			2.9	2.7	2.4	2.2	2.3	2.4	2.6	2.7	2.9	3.1	3.5	3.5				
	64.0			2.8	2.6	2.3	2.1	2.2	2.4	2.5	2.7	2.8	3.0	3.3	3.4				
	80.0			2.6	2.3	1.9	2.0	2.1	2.2	2.3	2.5	2.6	2.8	3.2					
	100.0		2.6	2.4	2.0	1.7	1.8	1.9	2.1	2.2	2.3	2.4	2.6	2.9					
	200.0	2.1	1.8	1.3	1.3	1.4	1.5	1.6	1.6	1.7	1.8	1.9		2.7					

AGMA Quality Number	Normal Diametral Pitch	0.040	0.063	0.100	0.160	0.250	0.400	0.630	1.0	1.6	2.5	4.0	6.3	10.	16.	25.	50.	100.	200.
	0.5																		
	1.0																		
	2.0																		
	4.0																		
	8.0																		
	12.0																		
	20.0						2.6	2.6	2.6	2.7	2.9	3.1	3.3	3.5	3.8	4.1	4.5		
	24.0						2.5	2.4	2.3	2.5	2.7	2.9	3.1	3.3	3.5	3.8	4.2		
	30.0					2.5	2.3	2.2	2.2	2.3	2.5	2.7	2.8	3.0	3.2	3.5			
Q15	32.0					2.4	2.3	2.1	2.1	2.3	2.4	2.6	2.8	3.0	3.2	3.4			
	40.0				2.4	2.2	2.0	1.9	2.0	2.1	2.2	2.4	2.6	2.7	2.9	3.1			
	48.0				2.3	2.1	1.8	1.8	1.9	2.0	2.1	2.2	2.4	2.5	2.7	2.9			
	50.0				2.2	2.0	1.7	1.7	1.8	2.0	2.1	2.2	2.4	2.5	2.7				
	60.0			2.3	2.1	1.8	1.5	1.6	1.7	1.8	2.0	2.1	2.2	2.3	2.5				
	64.0			2.2	2.0	1.7	1.5	1.6	1.7	1.8	1.9	2.0	2.2	2.3	2.4				
	80.0			2.1	1.8	1.4	1.4	1.5	1.6	1.7	1.8	1.9	2.0	2.1					
	100.0		2.1	1.9	1.5	1.2	1.3	1.4	1.5	1.6	1.6	1.7	1.8	1.9					
	200.0	1.8	1.4	0.9	1.0	1.0	1.1	1.1	1.2	1.2	1.3	1.4							

(continued)

197

TABLE 7.9 (continued)

AGMA Quality Number	Normal Diametral Pitch	0.040	0.063	0.100	0.160	0.250	0.400	0.630	1.0	1.6	2.5	4.0	6.3	10.	16.	25.	50.	100.	200.
	0.5																		
	1.0																		
	2.0																		
	4.0																		
	8.0																		
	12.0																		
	20.0						1.8	1.8	2.8	2.9	2.0	2.2	2.3	2.5	2.7	2.9	3.2		
	24.0					1.9	1.8	1.7	1.7	1.8	1.9	2.1	2.2	2.4	2.5	2.7	3.0		
	30.0					1.8	1.7	1.5	1.6	1.7	1.8	1.9	2.0	2.2	2.3	2.5			
	32.0					1.8	1.7	1.5	1.5	1.6	1.7	1.9	2.0	2.1	2.3	2.4			
	40.0				1.8	1.7	1.5	1.3	1.4	1.5	1.6	1.7	1.8	1.9	2.1	2.2			
Q16	48.0				1.8	1.6	1.3	1.3	1.3	1.4	1.5	1.6	1.7	1.8	1.9	2.1			
	50.0				1.7	1.5	1.2	1.2	1.3	1.4	1.5	1.6	1.7	1.8	1.9				
	60.0			1.8	1.6	1.4	1.1	1.2	1.2	1.3	1.4	1.5	1.6	1.7	1.8				
	64.0			1.8	1.6	1.3	1.1	1.1	1.2	1.3	1.4	1.5	1.5	1.6	1.7				
	80.0			1.7	1.4	1.0	1.0	1.1	1.1	1.2	1.3	1.3	1.4	1.5					
	100.0		1.7	1.5	1.1	0.9	0.9	1.0	1.1	1.1	1.2	1.2	1.3	1.4					
	200.0	1.5	1.2	0.7	0.7	0.7	0.8	0.8	0.8	0.9	0.9	1.0							

Source: Ref. 1.

TABLE 7.10 Tooth-to-Tooth Composite Tolerance (tolerances in ten-thousandths of an inch)

AGMA Quality Number	Normal Diametral Pitch	0.04	0.06	0.10	0.16	0.25	0.40	0.63	1.0	1.6	2.5	4.0	6.3	10.	16.	25.	50.	100.	200.
Q5	0.5																		
	1.0																		
	2.0																		
	4.0																		
	8.0																		
	12.0																		
	20.0						44.4	39.8	35.6	33.5	33.4	33.4	33.4	33.4	33.4	33.4	33.4		
	24.0					45.5	40.7	36.5	33.4	32.0	32.0	32.0	32.0	32.0	32.0	32.0	32.0		
	30.0					40.9	36.5	32.8	30.8	30.3	30.3	30.3	30.3	30.3	30.3	30.3			
	32.0					39.7	35.4	32.0	30.1	29.9	29.9	29.9	29.9	29.9	29.9	29.9			
	40.0				39.7	35.6	31.8	29.5	28.3	28.3	28.3	28.3	28.3	28.3	28.3	28.3			
	48.0				36.3	32.6	29.2	27.6	27.1	27.1	27.1	27.1	27.1	27.1	27.1	27.1			
	50.0				35.6	32.0	28.6	27.2	26.8	26.8	26.8	26.8	26.8	26.82	26.8				
	60.0			36.5	32.6	29.3	27.0	25.7	25.7	25.7	25.7	25.7	25.7	25.7	25.7				
	64.0			35.4	31.7	28.4	26.4	25.3	25.3	25.3	25.3	25.3	25.3	25.3	25.3				
	80.0			31.8	28.4	25.5	24.4	24.0	24.0	24.0	24.0	24.0	24.0	24.0					
	100.0		32.0	28.6	25.5	23.9	22.7	22.7	22.7	22.7	22.7	22.7	22.7	22.7					
	200.0	25.5	22.9	20.5	19.7	19.2	19.2	19.2	19.2	19.2	19.2	19.2							

(continued)

199

TABLE 7.10 (continued)

AGMA Quality Number	Normal Diametral Pitch	0.04	0.06	0.10	0.16	0.25	0.40	0.63	1.0	1.6	2.5	4.0	6.3	10.	16.	25.	50.	100.	200.
	0.5														76.9	69.1	57.8	57.9	57.9
	1.0												68.9	61.7	55.1	49.3	49.0	49.0	49.0
	2.0											55.1	49.4	44.2	40.7	41.5	41.5	41.5	41.5
	4.0									49.2	44.2	39.5	35.8	35.1	35.1	35.1	35.1	35.1	35.1
	8.0								39.5	35.3	31.7	29.6	29.7	29.7	29.7	29.7	29.7	29.7	
Q6	12.0							36.3	32.5	29.1	27.2	27.0	27.0	27.0	27.0	27.0	27.0	27.0	
	20.0						31.7	28.4	25.5	24.0	23.9	23.9	23.9	23.9	23.9	23.9	23.9		
	24.0					32.5	29.1	26.1	23.8	22.9	22.9	22.9	22.9	22.9	22.9	22.9	22.9		
	30.0					29.2	26.1	23.4	22.0	21.7	21.7	21.7	21.7	21.7	21.7	21.7			
	32.0					28.3	25.3	22.9	21.5	21.3	21.3	21.3	21.3	21.3	21.3	21.3			
	40.0				28.3	25.5	22.7	21.1	20.2	20.2	20.2	20.2	20.2	20.2	20.2	20.2			
	48.0				26.0	23.3	20.8	19.7	19.4	19.4	19.4	19.4	19.4	19.4	19.4	19.4			
	50.0				25.5	22.9	20.4	19.4	19.2	19.2	19.2	19.2	19.2	19.2	19.2				
	60.0			26.1	23.3	21.0	19.3	18.3	18.3	18.3	18.3	18.3	18.3	18.3	18.3				
	64.0			25.3	22.6	20.3	18.9	18.1	18.1	18.1	18.1	18.1	18.1	18.1	18.1				
	80.0			22.7	20.3	18.2	17.4	17.1	17.1	17.1	17.1	17.1	17.1	17.1					
	100.0		22.8	20.4	18.2	17.1	16.2	16.2	16.2	16.2	16.2	16.2	16.2	16.2					
	200.0	18.2	16.4	14.6	14.1	13.7	13.7	13.7	13.7	13.7	13.7	13.7							

AGMA Quality Number	Normal Diametral Pitch	0.04	0.06	0.10	0.16	0.25	0.40	0.63	1.0	1.6	2.5	4.0	6.3	10.	16.	25.	50.	100.	200.
Q7	0.5														54.9	49.3	41.3	41.3	41.3
	1.0												49.2	44.1	39.4	35.2	35.0	35.0	35.0
	2.0											39.4	35.3	31.6	29.1	29.6	29.6	29.6	29.6
	4.0									35.2	31.6	28.2	25.6	25.1	25.1	25.1	25.1	25.1	25.1
	8.0								28.2	25.2	22.7	21.1	21.2	21.2	21.2	21.2	21.2	21.2	
	12.0							26.0	23.2	20.8	19.4	19.3	19.3	19.3	19.3	19.3	19.3	19.3	
	20.0						22.7	20.3	18.2	17.1	17.1	17.1	17.1	17.1	17.1	17.1	17.1		
	24.0					23.2	20.8	18.6	17.0	16.3	16.3	16.3	16.3	16.3	16.3	16.3	16.3		
	30.0					20.9	18.6	16.7	15.7	15.5	15.5	15.5	15.5	15.5	15.5	15.5			
	32.0					20.2	18.1	16.3	15.4	15.2	15.2	15.2	15.2	15.2	15.2	15.2			
	40.0				20.2	18.2	16.2	15.0	15.4	14.4	14.4	14.4	14.4	14.4	14.4	14.4			
	48.0				18.5	16.7	14.9	14.1	13.8	13.8	13.8	13.8	13.8	13.8	13.8	13.8			
	50.0				18.2	16.3	14.6	13.9	13.7	13.7	13.7	13.7	13.7	13.7	13.7				
	60.0			18.6	16.7	15.0	13.8	13.1	13.1	13.1	13.1	13.1	13.1	13.1	13.1				
	64.0			18.1	16.1	14.5	13.5	12.9	12.9	12.9	12.9	12.9	12.9	12.9	12.9				
	80.0			16.2	14.5	13.0	12.4	12.2	12.2	12.2	12.2	12.2	12.2	12.2					
	100.0		16.3	14.6	13.0	12.2	11.6	11.6	11.6	11.6	11.6	11.6	11.6	11.6					
	200.0	13.0	11.7	10.5	10.1	9.8	9.8	9.8	9.8	9.8	9.8	9.8							

(continued)

201

TABLE 7.10 (continued)

AGMA Quality Number	Normal Diametral Pitch	0.04	0.06	0.10	0.16	0.25	0.40	0.63	1.0	1.6	2.5	4.0	6.3	10.	16.	25.	50.	100.	200.
Q8	0.5														39.2	35.2	29.5	29.5	29.5
	1.0												35.2	31.5	28.1	25.1	25.0	25.0	25.0
	2.0											28.1	25.2	22.6	20.8	21.2	21.2	21.2	21.2
	4.0									25.1	22.6	20.2	18.3	17.9	17.9	17.9	17.9	17.9	17.9
	8.0								20.2	18.0	16.2	15.1	15.2	15.2	15.2	15.2	15.2	15.2	
	12.0							18.5	16.6	14.8	13.9	13.8	13.8	13.8	13.8	13.8	13.8	13.8	
	20.0						16.2	14.5	13.0	12.2	12.2	12.2	12.2	12.2	12.2	12.2	12.2		
	24.0					16.6	14.8	13.3	12.2	11.7	11.7	11.7	11.7	11.7	11.7	11.7	11.7		
	30.0					14.9	13.3	11.9	11.2	11.1	11.1	11.1	11.1	11.1	11.1	11.1			
	32.0					14.5	12.9	11.6	11.0	10.9	10.9	10.9	10.9	10.9	10.9	10.9			
	40.0				14.5	13.0	11.6	10.7	10.3	10.3	10.3	10.3	10.3	10.3	10.3	10.3			
	48.0				13.2	11.9	10.6	10.1	9.9	9.9	9.9	9.9	9.9	9.9	9.9	9.9			
	50.0				13.0	11.7	10.4	9.9	9.8	9.8	9.8	9.8	9.8	9.8	9.8				
	60.0			13.3	11.9	10.7	9.9	9.4	9.4	9.4	9.4	9.4	9.4	9.4	9.4				
	64.0			12.9	11.5	10.4	9.6	9.2	9.2	9.2	9.2	9.2	9.2	9.2	9.2				
	80.0			11.6	10.4	9.3	8.9	8.7	8.7	8.7	8.7	8.7	8.7	8.7					
	100.0		11.6	10.4	9.3	8.7	8.3	8.3	8.3	8.3	8.3	8.3	8.3	8.3					
	200.0	9.3	8.3	7.5	7.2	7.0	7.0	7.0	7.0	7.0	7.0	7.0							

AGMA Quality Number	Normal Diametral Pitch	0.04	0.06	0.10	0.16	0.25	0.40	0.63	1.0	1.6	2.5	4.0	6.3	10.	16.	25.	50.	100.	200.
Q9	0.5														28.0	25.2	21.1	21.1	21.1
	1.0												25.1	22.5	20.1	18.0	17.9	17.9	17.9
	2.0											20.1	18.0	16.1	14.8	15.1	15.1	15.1	15.1
	4.0									17.9	16.1	14.4	13.0	12.8	12.8	12.8	12.8	12.8	12.8
	8.0								14.4	12.9	11.6	10.8	10.8	10.8	10.8	10.8	10.8	10.8	
	12.0							13.2	11.9	10.6	9.9	9.8	9.8	9.8	9.8	9.8	9.8	9.8	
	20.0					11.9	11.6	10.4	9.3	8.7	8.7	8.7	8.7	8.7	8.7	8.7	8.7		
	24.0					10.6	10.6	9.5	8.7	8.3	8.3	8.3	8.3	8.3	8.3	8.3	8.3		
	30.0						9.5	8.5	8.0	7.9	7.9	7.9	7.9	7.9	7.9	7.9			
	32.0					10.3	9.2	8.3	7.8	7.8	7.8	7.8	7.8	7.8	7.8	7.8			
	40.0				10.3	9.3	8.3	7.7	7.4	7.4	7.4	7.4	7.4	7.4	7.4	7.4			
	48.0				9.5	8.5	7.6	7.2	7.1	7.1	7.1	7.1	7.1	7.1	7.1	7.1			
	50.0				9.3	8.3	7.4	7.1	7.0	7.0	7.0	7.0	7.0	7.0	7.0				
	60.0			9.5	8.5	7.6	7.0	6.7	6.7	6.7	6.7	6.7	6.7	6.7	6.7				
	64.0			9.2	8.2	7.4	6.9	6.6	6.6	6.6	6.6	6.6	6.6	6.6	6.6				
	80.0			8.3	7.4	6.7	6.3	6.2	6.2	6.2	6.2	6.2	6.2	6.2					
	100.0		8.3	7.4	6.7	6.2	5.9	5.9	5.9	5.9	5.9	5.9	5.9	5.9					
	200.0	6.7	6.0	5.3	5.1	5.0	5.0	5.0	5.0	5.0	5.0	5.0							

(continued)

TABLE 7.10 (continued)

AGMA Quality Number: Q10

Normal Diametral Pitch	0.04	0.06	0.10	0.16	0.25	0.40	0.63	1.0	1.6	2.5	4.0	6.3	10.	16.	25.	50.	100.	200.
0.5														20.0	18.0	15.0	15.1	15.1
1.0												17.9	16.1	14.3	12.8	12.8	12.8	12.8
2.0											14.3	12.9	11.5	10.6	10.8	10.8	10.8	10.8
4.0									12.8	11.5	10.3	9.3	9.1	9.1	9.1	9.1	9.1	9.1
8.0								10.3	9.2	8.3	7.7	7.7	7.7	7.7	7.7	7.7	7.7	
12.0							9.5	8.5	7.6	7.1	7.0	7.0	7.0	7.0	7.0	7.0	7.0	
20.0						8.3	7.4	6.6	6.2	6.2	6.2	6.2	6.2	6.2	6.2	6.2		
24.0					8.5	7.6	6.8	6.2	5.9	5.9	5.9	5.9	5.9	5.9	5.9	5.9		
30.0					7.6	6.8	6.1	5.7	5.6	5.6	5.6	5.6	5.6	5.6	5.6			
32.0					7.4	6.6	5.9	5.6	5.6	5.6	5.6	5.6	5.6	5.6	5.6			
40.0				7.4	6.6	5.9	5.5	5.3	5.3	5.3	5.3	5.3	5.3	5.3	5.3			
48.0				6.8	6.1	5.4	5.1	5.0	5.0	5.0	5.0	5.0	5.0	5.0	5.0			
50.0				6.6	6.0	5.3	5.1	5.0	5.0	5.0	5.0	5.0	5.0	5.0				
60.0			6.8	6.1	5.5	5.0	4.8	4.8	4.8	4.8	4.8	4.8	4.8	4.8				
64.0			6.6	5.9	5.3	4.9	4.7	4.7	4.7	4.7	4.7	4.7	4.7	4.7				
80.0			5.9	5.3	4.8	4.5	4.5	4.5	4.5	4.5	4.5	4.5	4.5					
100.0		5.9	5.3	4.8	4.4	4.2	4.2	4.2	4.2	4.2	4.2	4.2	4.2					
200.0	4.8	4.3	3.8	3.7	3.6	3.6	3.6	3.6	3.6	3.6	3.6							

AGMA Quality Number	Normal Diametral Pitch	0.04	0.06	0.10	0.16	0.25	0.40	0.63	1.0	1.6	2.5	4.0	6.3	10.	16.	25.	50.	100.	200.
Q11	0.5														14.3	12.8	10.7	10.8	10.8
	1.0												12.8	11.5	10.2	9.2	9.1	9.1	9.1
	2.0											10.2	9.2	8.2	7.6	7.7	7.7	7.7	7.7
	4.0									9.2	8.2	7.3	6.7	6.5	6.5	6.5	6.5	6.5	7.7
	8.0								7.3	6.6	5.9	5.5	5.5	5.5	5.5	5.5	5.5	5.5	6.5
	12.0							6.8	6.0	5.4	5.1	5.0	5.0	5.0	5.0	5.0	5.0	5.0	
	20.0						5.9	5.3	4.7	4.5	4.4	4.4	4.4	4.4	4.4	4.4	4.4		
	24.0					6.0	5.4	4.8	4.4	4.2	4.2	4.2	4.2	4.2	4.2	4.2	4.2		
	30.0					5.4	4.9	4.4	4.1	4.0	4.0	4.0	4.0	4.0	4.0	4.0			
	32.0					5.3	4.7	4.2	4.0	4.0	4.0	4.0	4.0	4.0	4.0	4.0			
	40.0				5.3	4.7	4.2	3.9	3.8	3.8	3.8	3.8	3.8	3.8	3.8	3.8			
	48.0				4.8	4.3	3.9	3.7	3.6	3.6	3.6	3.6	3.6	3.6	3.6	3.6			
	50.0				4.7	4.3	3.8	3.6	3.6	3.6	3.6	3.6	3.6	3.6	3.6				
	60.0			4.9	4.3	3.9	3.6	3.4	3.4	3.4	3.4	3.4	3.4	3.4	3.4				
	64.0			4.7	4.2	3.8	3.5	3.4	3.4	3.4	3.4	3.4	3.4	3.4	3.4				
	80.0			4.2	3.8	3.4	3.2	3.2	3.2	3.2	3.2	3.2	3.2	3.2					
	100.0		4.2	3.8	3.4	3.2	3.0	3.0	3.0	3.0	3.0	3.0	3.0	3.0					
	200.0	3.4	3.0	2.7	2.6	2.6	2.6	2.6	2.6	2.6	2.6	2.6							

(continued)

TABLE 7.10 (continued)

AGMA Quality Number	Normal Diametral Pitch	0.04	0.06	0.10	0.16	0.25	0.40	0.63	1.0	1.6	2.5	4.0	6.3	10.	16.	25.	50.	100.	200.
	0.5														10.2	9.2	7.7	7.7	7.7
	1.0												9.2	8.2	7.3	6.5	6.5	6.5	6.5
	2.0									6.5	5.9	7.3	6.6	5.9	5.4	5.5	5.5	5.5	5.5
	4.0								5.2	4.7	4.2	5.2	4.8	4.7	4.7	4.7	4.7	4.7	4.7
	8.0								4.3	3.9	3.6	3.9	4.0	4.0	4.0	4.0	4.0	4.0	
Q12	12.0							4.8	3.4	3.2	3.2	3.6	3.6	3.6	3.6	3.6	3.6	3.6	
	20.0						4.2	3.8	3.4	3.0	3.2	3.2	3.2	3.2	3.2	3.2	3.2		
	24.0					4.3	3.9	3.5	3.2	2.9	3.0	3.0	3.0	3.0	3.0	3.0	3.0		
	30.0					3.9	3.5	3.1	2.9	2.9	2.9	2.9	2.9	2.9	2.9	2.9			
	32.0					3.8	3.4	3.0	2.9	2.8	2.8	2.8	2.8	2.8	2.8	2.8			
	40.0				3.8	3.4	3.0	2.8	2.7	2.7	2.7	2.7	2.7	2.7	2.7	2.7			
	48.0				3.4	3.1	2.8	2.6	2.6	2.6	2.6	2.6	2.6	2.6	2.6	2.6			
	50.0				3.4	3.0	2.7	2.6	2.5	2.5	2.5	2.5	2.5	2.5	2.5				
	60.0			3.5	3.1	2.8	2.6	2.4	2.4	2.4	2.4	2.4	2.4	2.4	2.4				
	64.0			3.4	3.0	2.7	2.5	2.4	2.4	2.4	2.4	2.4	2.4	2.4	2.4				
	80.0		3.0	3.0	2.7	2.4	2.3	2.3	2.3	2.3	2.3	2.3	2.3	2.3					
	100.0			2.7	2.4	2.3	2.2	2.2	2.2	2.2	2.2	2.2	2.2	2.2					
	200.0	2.4	2.2	1.9	1.9	1.8	1.8	1.8	1.8	1.8	1.8	1.8							

Table (AGMA Quality Number Q13)

AGMA Quality Number	Normal Diametral Pitch	0.04	0.06	0.10	0.16	0.25	0.40	0.63	1.0	1.6	2.5	4.0	6.3	10.	16.	25.	50.	100.	200.
	0.5														7.3	6.6	5.5	5.5	5.5
	1.0												6.5	5.9	5.2	4.7	4.6	4.6	4.6
	2.0											5.2	4.7	4.2	3.9	3.9	3.9	3.9	3.9
	4.0							3.4	3.7	4.7	4.2	3.7	3.4	3.3	3.3	3.3	3.3	3.3	3.3
	8.0						3.0	2.7	3.1	3.3	3.0	2.8	2.8	2.8	2.8	2.8	2.8	2.8	
Q13	12.0					3.1	2.8	2.5	2.4	2.8	2.6	2.6	2.6	2.6	2.6	2.6	2.6	2.6	
	20.0					2.8	2.5	2.2	2.3	2.3	2.3	2.3	2.3	2.3	2.3	2.3	2.3		
	24.0				2.7	2.7	2.4	2.2	2.1	2.2	2.2	2.2	2.2	2.2	2.2	2.2	2.2		
	30.0				2.5	2.4	2.2	2.0	2.0	2.1	2.1	2.1	2.1	2.1	2.1	2.1			
	32.0			2.5	2.4	2.2	2.0	1.9	1.9	2.0	2.0	2.0	2.0	2.0	2.0	2.0			
	40.0			2.4	2.2	2.0	1.9	1.8	1.8	1.9	1.9	1.9	1.9	1.9	1.9	1.9			
	48.0			2.2	2.1	2.0	1.8	1.7	1.8	1.8	1.8	1.8	1.8	1.8	1.8	1.8			
	50.0			1.9	1.9	2.0	1.8	1.7	1.7	1.8	1.8	1.8	1.8	1.8	1.8				
	60.0			1.4	1.7	1.9	1.7	1.6	1.7	1.7	1.7	1.7	1.7	1.7	1.7				
	64.0				1.3	1.7	1.5	1.5	1.6	1.7	1.7	1.7	1.7	1.7	1.7				
	80.0		2.2			1.3	1.3	1.3	1.5	1.6	1.6	1.6	1.6	1.6					
	100.0		1.6						1.3	1.5	1.5	1.5	1.5						
	200.0	1.7								1.3	1.3	1.3							

(continued)

TABLE 7.10 (continued)

AGMA Quality Number	Normal Diametral Pitch	0.04	0.06	0.10	0.16	0.25	0.40	0.63	1.0	1.6	2.5	4.0	6.3	10.	16.	25.	50.	100.	200.
	0.5														5.2	4.7	3.9	3.9	3.9
	1.0												4.7	4.2	3.7	3.3	3.3	3.3	3.3
	2.0									3.3	3.0	3.7	3.3	3.0	2.8	2.8	2.8	2.8	2.8
	4.0								2.7	2.4	2.1	2.7	2.4	2.4	2.4	2.4	2.4	2.4	2.4
	8.0								2.2	2.0	1.8	2.0	2.0	2.0	2.0	2.0	2.0	2.0	
Q14	12.0							2.5	1.7	1.6	1.6	1.8	1.8	1.8	1.8	1.8	1.8	1.8	
	20.0						2.1	1.9	1.6	1.5	1.5	1.6	1.6	1.6	1.6	1.6	1.6		
	24.0					2.2	2.0	1.8	1.5	1.5	1.5	1.5	1.5	1.5	1.5	1.5	1.5		
	30.0					2.0	1.8	1.6	1.5	1.5	1.5	1.5	1.5	1.5	1.5	1.5			
	32.0					1.9	1.7	1.5	1.4	1.4	1.4	1.4	1.4	1.4	1.4	1.4			
	40.0				1.9	1.7	1.5	1.4	1.3	1.4	1.4	1.4	1.4	1.4	1.4	1.4			
	48.0				1.8	1.6	1.4	1.3	1.3	1.3	1.3	1.3	1.3	1.3	1.3	1.3			
	50.0				1.7	1.5	1.4	1.3	1.3	1.3	1.3	1.3	1.3	1.3	1.3				
	60.0			1.8	1.6	1.4	1.3	1.2	1.2	1.2	1.2	1.2	1.2	1.2	1.2				
	64.0			1.7	1.5	1.4	1.3	1.2	1.2	1.2	1.2	1.2	1.2	1.2	1.2				
	80.0			1.5	1.4	1.2	1.2	1.2	1.2	1.2	1.2	1.2	1.2	1.2					
	100.0		1.5	1.4	1.2	1.2	1.1	1.1	1.1	1.1	1.1	1.1	1.1	1.1					
	200.0	1.2	1.1	1.0	1.0	0.9	0.9	0.9	0.9	0.9	0.9	0.9							

AGMA Quality Number	Normal Diametral Pitch	0.04	0.06	0.10	0.16	0.25	0.40	0.63	1.0	1.6	2.5	4.0	6.3	10.	16.	25.	50.	100.	200.
	0.5																		
	1.0																		
	2.0																		
Q15	4.0																		
	8.0																		
	12.0																		
	20.0						1.5	1.4	1.2	1.2	1.2	1.2	1.2	1.2	1.2	1.2	1.2		
	24.0					1.6	1.4	1.3	1.2	1.1	1.1	1.1	1.1	1.1	1.1	1.1	1.1		
	30.0					1.4	1.3	1.1	1.1	1.0	1.0	1.0	1.0	1.0	1.0	1.0			
	32.0					1.4	1.2	1.1	1.0	1.0	1.0	1.0	1.0	1.0	1.0	1.0			
	40.0				1.4	1.2	1.1	1.0	1.0	1.0	1.0	1.0	1.0	1.0	1.0	1.0			
	48.0				1.3	1.1	1.0	1.0	0.9	0.9	0.9	0.9	0.9	0.9	0.9	0.9			
	50.0				1.2	1.1	1.0	0.9	0.9	0.9	0.9	0.9	0.9	0.9	0.9				
	60.0			1.3	1.1	1.0	0.9	0.9	0.9	0.9	0.9	0.9	0.9	0.9	0.9				
	64.0			1.2	1.1	1.0	0.9	0.9	0.9	0.9	0.9	0.9	0.9	0.9	0.9				
	80.0			1.1	1.0	0.9	0.8	0.8	0.8	0.8	0.8	0.8	0.8	0.8					
	100.0		1.1	1.0	0.9	0.8	0.8	0.8	0.8	0.8	0.8	0.8	0.8	0.8					
	200.0	0.9	0.8	0.7	0.7	0.7	0.7	0.7	0.7	0.7	0.7	0.7							

(continued)

TABLE 7.10 (continued)

AGMA Quality Number	Normal Diametral Pitch	0.04	0.06	0.10	0.16	0.25	0.40	0.63	1.0	1.6	2.5	4.0	6.3	10.	16.	25.	50.	100.	200.
	0.5																		
	1.0																		
	2.0																		
	4.0																		
	8.0																		
Q16	12.0							1.0	0.9	0.8	0.8	0.8	0.8	0.8	0.8	0.8	0.8		
	20.0						1.1	0.9	0.8	0.8	0.8	0.8	0.8	0.8	0.8	0.8	0.8		
	24.0					1.1	1.0	0.8	0.8	0.7	0.7	0.7	0.7	0.7	0.7	0.7			
	30.0					1.0	0.9	0.8	0.7	0.7	0.7	0.7	0.7	0.7	0.7	0.7			
	32.0					1.0	0.9	0.7	0.7	0.7	0.7	0.7	0.7	0.7	0.7	0.7			
	40.0				1.0	0.9	0.8	0.7	0.7	0.7	0.7	0.7	0.7	0.7	0.7	0.7			
	48.0				0.9	0.8	0.7	0.7	0.7	0.7	0.7	0.7	0.7	0.7	0.7				
	50.0				0.9	0.8	0.7	0.6	0.6	0.6	0.6	0.6	0.6	0.6	0.6				
	60.0			0.9	0.8	0.7	0.7	0.6	0.6	0.6	0.6	0.6	0.6	0.6	0.6				
	64.0			0.9	0.8	0.7	0.7	0.6	0.6	0.6	0.6	0.6	0.6	0.6					
	80.0			0.8	0.7	0.6	0.6	0.6	0.6	0.6	0.6	0.6	0.6						
	100.0		0.8	0.7	0.6	0.6	0.6	0.6	0.6	0.6	0.6	0.6							
	200.0	0.6	0.6	0.5	0.5	0.5	0.5	0.5	0.5	0.5	0.5	0.5							

Source: Ref. 1.

TABLE 7.11 Rack Tolerances, Coarse- and Fine-Pitch (in ten-thousandths of an inch)

AGMA QUALITY NUMBER	NORMAL DIAMETRAL PITCH	PITCH TOLERANCE plus or minus	INDEX TOLERANCE per foot of rack length plus or minus	Pitch Plane to reference surface tolerance in any 12" (incl. variation across face) minus FACE WIDTH, INCHES					Total pitch plane to reference surface tolerance in 72" (incl. variation across face) minus FACE WIDTH, INCHES				
				up to 1	1-2.49	2.5-3.9	4-7.9	8-12	up to 1	1-2.49	2.5-3.9	4-7.9	8-12
Q3	1/2	120	360										
	1	99	297										
	2	88	264										
	4	74	222										
Q4	1/2	86	258										
	1	70	210										
	2	63	189										
	4	53	159										
Q5	1/2	61	183										
	1	50	150										
	2	45	135										
	4	38	114										
	8	32	96										
Q6	1/2	44	132									480	600
	1	36	108									360	480
	2	32	96								280	320	400
	4	27	81							180	240		
	8	23	69						130	150			
	16-19.9	20	60						110				
Q7	1/2	31	93									350	400
	1	26	78									280	320
	2	23	69								220	250	300
	4	19	57							150	180		
	8	17	51						100	115			
	16-19.9	15	45						90				
	20-48	14	42						75				

(continued)

Outside diameter of the gears determines the addendum and has a direct influence on the contact ratio. The tops of gear teeth are sometimes used for a locating surface for subsequent machining operations. Although an uncommon practice in plastics gearing, the outside diameter is particularly important if such an operation is used in manufacturing. Tolerance on the outside diameter has been standardized to satisfy functional requirements. To ensure adequate clearance, the tolerance is specified all minus. Table 7.6 provides tolerance values for both diametral pitch and metric module gears.

Tolerance values of Table 7.7 are tooth thickness quality classes for spur, helical, herringbone, rack, and pinions. These are AGMA classes when tooth thickness is specified with a tolerance by classification letter. Tooth thickness must be inspected in cases in which

TABLE 7.11 (continued)

AGMA QUALITY NUMBER	NORMAL DIAMETRAL PITCH	PITCH TOLERANCE plus or minus	INDEX TOLERANCE per foot of rack length plus or minus	Pitch plane to reference surface tolerance in any 12" (incl. variation across face) minus — FACE WIDTH, INCHES					Total pitch plane to reference surface tolerance in 72" (incl. variation across face) minus — FACE WIDTH, INCHES				
				up to 1	1-2.49	2.5-3.9	4-7.9	8-12	up to 1	1-2.49	2.5-3.9	4-7.9	8-12
Q8	1	18	54									225	250
	2	16	48								170	200	230
	4	14	42							110	140		
	8	12	36						80	90			
	16-19.9	10	30						70				
	20-120	8	24						60				
Q9	1	13	39				90	105				180	210
	2	12	36			65	80	95			135	160	190
	4	10	29		45	55				90	115		
	8	8	25	30	35				65	75			
	16-19.9	7	22	25					55				
	20-120	6	17	20					45				
Q10	1	9	28				75	90				150	180
	2	8	25			50	60	75			105	125	150
	4	7	22		35	45				70	90		
	8	6	18	25	30				50	60			
	16-19.9	5	16	20					45				
	20-120	4	12	15					35				
Q11	1	7	20				60	75				120	150
	2	6	18			35	45	60			75	95	120
	4	5	15		25	30				55	65		
	8	4	13	20	22				40	45			
	16-19.9	4	11	15					35				
	20-120	3	9	13					25				
Q12	2	4	13			28	35	48			55	70	95
	4	4	11		20	25				40	50		
	8	3	9	15	17				30	35			
	16-19.9	3	8	13					25				
	20-120	2	6	10									
Q13	2	3	9			23	25	38			45	50	75
	4	3	8		15	20				30	40		
	8	2	7	12	13				24	27			
	16-19.9	2	6	10					20				
	20-120	2	5	7									
Q14	2	2	7			17	20	30			35	40	60
	4	2	5		13	15				25	30		
	8	2	5	10	12				20	23			
	16-19.9	2	5	7					15				
	20-120	2	5	5									

tolerances are specified for tooth elements instead of inspected using a running-gage, as explained in Chapter 10. Table 7.8 provides additional tolerances for runout, pitch, profile, and lead.

Composite Tolerances

In contrast to individual element inspection, Tables 7.9 and 7.10 are tolerances for total composite and tooth-to-tooth inspection. Table

TOOTH-TO-TOOTH COMPOSITE TOLERANCE	TOTAL COMPOSITE TOLERANCE in any 12" (1½" face max)	TOTAL COMPOSITE TOLERANCE in 72" (1½" face max)	PROFILE TOLERANCE (see note 11 page 45)	LEAD TOLERANCE FACE WIDTH, INCH				
				To 1"	2	3	4	5
			22					
			18					
			13	4	7	9	11	13
11	31	62	10					
9	25	50	8					
7	20	40						
			16					
			13					
			10	3	5	7	9	10
8	22	44	7					
6	18	36	6					
5	14	28						
			11					
			9					
			7	3	4	6	7	8
6	16	32	5					
5	13	26	4					
4	10	20						
			7					
			5			3	5	6
4	11	22	4					
3	9	18	3					
3	7							
			4					
			4	2	3	4	5	5
3	8	16	3					
2	7		2					
2	5							
			3					
			3	1	2	3	4	4
2	6	12	2					
2	5		2					
1	4							

Note: Classes Q12, Q13, and Q14 in the shaded area
are only recommended for highly specialized applications.
Source: Ref. 1.

7.11 for coarse- and fine-pitch racks provides pitch, index, mounting
location, total composite, tooth-to-tooth composite, and lead tolerances.
Values may be interpolated for gear diameters and pitches falling between those listed. Appendix D is provided for computer computation
of runout, pitch, profile, tooth-to-tooth composite, and total composite tolerances.

TABLE 7.12 Bevel and Hypoid Tolerances (in ten-thousandths of an inch)

AGMA QUALITY NUMBER	DIA-METRAL PITCH	RUNOUT TOLERANCE PITCH DIAMETER (INCHES)									PITCH TOLERANCE (see para. 4., page 55)								
		3/4	1½	3	6	12	25	50	100	200	3/4	1½	3	6	12	25	50	100	200
Q3	½						770	1010	1360										
	1					540	710	930	1250										
	2				382	498	660	860	1150										
	4			280	355	460	608	800											
Q4	½						540	700	940										
	1					378	496	640	860										
	2				272	348	452	590	790										
	4			198	250	320	419	542	720										
Q5	¼						396	510	665	880									
	1					270	350	450	582	775									
	2				184	233	302	390	510	680									
	4			130	160	203	262	340	440	590									
	8		91	112	140	177	228	290	380										
	16-19,99																		
	20-24																		
	24-32																		
	32-48																		
Q6	½						280	350	450	600						50	55	62	70
	1					188	235	295	378	508					31	33	38	42	50
	2			131	160	200	250	322	425				26	27	29	33	37	42
	4			92	110	135	170	210	270	360			22	22	24	26	28	32	37
	8		64	76	93	114	143	180	230			18	18	19	21	23	25	28	
	16-19,99	46	55	66	80	98	122	152	193		16	16	16	17	18	19	21	23	
	20-32																		
	32-48																		
	48-64																		
Q7	½						209	260	335	445						37	40	45	50
	1					132	165	205	261	350					23	25	28	32	37
	2			84	103	130	165	210	280				19	20	22	24	27	31
	4			55	67	82	103	130	167	225			16	16	17	19	21	24	28
	8		37	44	54	66	82	103	132			13½	14	14½	15½	17	19	21	
	16-19,99	26	31	37	45	55	69	86	110		11	11½	12	12½	13	14	15½	17½	
	20-48																		
	48-64																		
	64-96																		

In Tables 7.8 through 7.11, some points to be noted are:

1. When specifying values from the tables on a print, the upper case "Q" should precede the numerical quality value; for example, "Q6."

2. Five-decimal-place accuracy is shown for aid in interpolation. It is suggested that values be rounded to four places for use on drawings or specifications.

TOOTH-TO-TOOTH COMPOSITE TOLERANCE							TOTAL COMPOSITE TOLERANCE						
PITCH DIAMETER (INCHES)							PITCH DIAMETER (INCHES)						
3/16	3/8	3/4	1½	3	6	12	3/16	3/8	3/4	1½	3	6	12
27	27	27	27	27	27	27	52	52	52	52	61	72	72
27	27	27	27	27			52	52	52	52	61		
27	27	27	27				52	52	52	52			
19	19	19	19	19	19	19	37	37	37	37	44	52	52
19	19	19	19	19			37	37	37	37	44		
19	19	19	19				37	37	37	37			
14	14	14	14	14	14	14	27	27	27	27	32	37	37
14	14	14	14	14			27	27	27	27	32		
14	14	14	14				27	27	27	27			

(continued)

3. Profile tolerance is exclusive of root and tip modification. Values are to be interpreted as maximum total deviation from high point to low point.
4. Lead tolerance is limited to 5.0-inch face width and is measured normal to axis of the tooth. Values are to be interpreted as maximum total deviation from high point to low point.

TABLE 7.12 (continued)

AGMA QUALITY NUMBER	DIAMETRAL PITCH	RUNOUT TOLERANCE									PITCH TOLERANCE								
		PITCH DIAMETER (INCHES)																	
		3/4	1½	3	6	12	25	50	100	200	3/4	1½	3	6	12	25	50	100	200
Q8	½						160	200	255							26	28	31	
	1					95	115	140	180							16	18	19	22
	2				58	68	82	100	125						14	14	15	17	19
	4			36	41	47	56	67						11	11	12	13	15	
	8		25	28	32	36	42						9	10	10	11	12		
	16-19,99	19	21	26	30						8	8	8	9	9				
	20-64																		
	64-96																		
	96-120																		
Q9	½						113	140	180							19	20	22	
	1					68	81	100	125							11	12	14	16
	2				40	48	58	70	88						10	10	11	12	14
	4			26	29	34	40	48						8	8	9	9½	11	
	8		18	20	22	26	30						7	7	7½	8	8½		
	16-19,99	14	15	16	18	21					6	6	6	6½	7				
	20-120																		
Q10	1					50	58									8½	9		
	2				30	34	40								7	7½	8		
	4			18	21	24	28							6	6	6½	7		
	8		13	14	16	18	21						5	5	5½	6	6½		
	16-19,99	10	11	12	13	15					4½	4½	4½	4½	5				
	20-120																		
Q11	1					34	41									6	6		
	2				21	24	28								5	5	6		
	4			13	15	17	20							4	4	4½	5		
	8		9	10	11	13	15						3½	3½	4	4	4½		
	16-19.99	7	8	8	9	11					3	3	3	3½	3½				
	20-120																		
Q12	2				15	18	21								3½	4	4		
	4			9	11	12	14							3	3	3½	3½		
	8		6½	7	8	9	11						2½	2½	3	3	3½		
	16-19.99	5½	6	7	8						2½	2½	2½	2½	2½				
	20-120																		
Q13	2				11	13	15								2½	3	3		
	4			7	7½	9	10½							2	2	2½	2½		
	8		5	5½	6	7	8						2	2	2	2	2½		
	16-19.99	5	5	5	5	5½					2	2	2	2	2				
	20-120																		

5. Pitch tolerance is given for the plane of rotation. If measured in the normal plane, the measured values should be divided by the cosine of the helical angle.

6. The critical accuracy application requirements of quality numbers Q13, Q14, Q15, and Q16 require agreement between the manufacturer and user as to the method of inspection.

Tables 7.12 through 7.14 show tolerances for bevel, hypoid, and fine-pitch worms and wormgearing. Table 7.12 provides runout,

TOOTH-TO-TOOTH COMPOSITE TOLERANCE							TOTAL COMPOSITE TOLERANCE						
PITCH DIAMETER (INCHES)							PITCH DIAMETER (INCHES)						
3/16	3/8	3/4	1½	3	6	12	3/16	3/8	3/4	1½	3	6	12
													80
											46	52	58
										35	38	42	46
										27	30	34	37
10	10	10	10	10	10	10	19	19	19	19	23	27	27
10	10	10	10	10			19	19	19	19	23		
10	10	10	10				19	19	19	19			
													57
											33	37	42
										24	27	30	33
										19	22	24	27
7	7	7	7	7	7		14	14	14	14	16	19	
													40
											23	26	29
										17	19	21	24
										14	15	17	19
5	5	5	5	5	5		10	10	10	10	12	14	
											17	18	21
										13	14	15	17
										10	11	12	14
4	4	4	4	4	4		7	7	7	7	9	10	
											12	13	15
										9	10	11	12
										7	8	9	10
3	3	3	3	3	3		5	5	5	5	6	7	
											10	11	12
										8	9	10	11
										6	7	8	
2	2	2	2	2	2		4	4	4	4	4	5	

Source: Ref. 1.

pitch, tooth-to-tooth, and composite tolerances for bevel and hypoid gears that usually require special fixtures for inspection. If it is necessary to calculate spacing tolerance, multiply the values of pitch tolerance by 1.5. If values of total index tolerance are necessary, multiply the values of pitch tolerance by 3.0.

TABLE 7.13 Fine-Pitch Worm and Wormgear Tolerances Axial Pitch
0.030 to 0.160 in. (in ten-thousandths of an inch)

AGMA quality numbers	Worm tolerances		Tooth-to-tooth tolerance[a]	Total composite tolerance
	Axial pitch tolerance	Lead tolerance[a]		
Q5	13	26	27	52
Q6	9	18	19	37
Q7	6	13	14	27
Q8	5	9	10	19
Q9	4	7	7	14
Q10	3	5	5	10
Q11	2	4	4	7
Q12	1.3	2.5	3	5
Q13	1.0	2.0	2	4
Q14	0.7	1.3	1.4	2.7
Q15	0.5	1.0	1.0	2.0
Q16	0.3	0.7	0.7	1.4

[a]Use axial pitch tolerances for lead tolerance of single thread worms.
Source: Ref. 1.

Assembly

Tolerance on standard center distance using the standard-center-distance system is also related to the AGMA quality number, as given in
Table 7.15.

The relationship of parallel shafts must be held within certain limits to provide proper gear performance. For the average spur gear
application using Q5 through Q11 gears, the axis relationship must
be held, as given in Table 7.16. The composite total of axial play,
lateral runout, and face misalignment of a pair of spur gears should
not exceed 10% of the planned face engagement.

TABLE 7.14 Tooth Thickness Tolerance Classes for Fine-Pitch Worms and Wormgearing (in ten-thousandths of an inch)

Quality number	Axial pitch (in.)	Diametral pitch	Class A	B	C	D	E
	0.160	20	16-40	8-20			
	0.130	24	15-35	7.6-18			
	0.100	31	14-30	7-15			
Q5	0.080	39	13-27	6.5-13.5			
	0.065	48	12-24	6-12			
	0.050	63	11-21	5.7-10.6			
	0.040	78	11-19	5.5-10			
	0.030	104	10-17	5.2-8.8			
	0.160	20	16-40	8-20	4-10		
	0.130	24	15-35	7.6-18	3.8-9		
	0.100	31	14-30	7-15	3.5-7.8		
Q6	0.080	39	13-27	6.5-13	3.3-6.8		
	0.065	48	12-24	6-12	3-6		
	0.050	63	11-21	5.7-10.6	2.9-5.4		
	0.040	78	11-19	5.5-10	2.8-5		
	0.030	104	10-17	5.2-8.8	2.6-4.4		
	0.160	20	16-40	8-20	4-10	2-5	1.1-2.7
	0.130	24	15-35	7.6-18	3.8-9	1.9-4.6	0-2.4
	0.100	31	14-30	7-15	3.5-7.8	1.8-3.8	0-2
Q7 and higher	0.080	39	13-27	6.5-13.5	3.3-6.8	1.7-3.5	0-1.8
	0.065	48	12-24	6-12	3-6	1.6-3	0-1.6
	0.050	63	11-21	5.7-10.6	2.9-5.4	1.5-2.7	0-1.4
	0.040	78	11-19	5.5-10	2.8-5	1.4-2.5	0-1.3
	0.030	104	10-17	5.2-8.8	2.6-4.4	1.3-2.3	0-1.2

Source: Ref. 1.

TABLE 7.15 Tolerance on Standard Center Distance
(plus or minus, in.)

AGMA quality number	Tolerance on standard center distance	AGMA quality number	Tolerance on standard center distance
Q5	0.0025	Q11	0.0007
Q6	0.0020	Q12	0.0006
Q7	0.0015	Q13	0.0005
Q8	0.0010	Q14	0.0004
Q9	0.0009	Q15	0.0003
Q10	0.0008	Q16	0.0002

TABLE 7.16 Shaft Parallelism

Face width (in.)	Parallel within in./in.
<0.375	0.0010
0.375-0.625	0.0007
>0.625	0.0005

Source: Ref. 3.

References

1. *AGMA Gear Handbook, Volume 1: Gear Classification, Materials and Measuring Methods for Unassembled Gears* (AGMA 390.03), American Gear Manufacturers Association, Alexandria, VA (1973).

2. *AGMA Design Manual for Fine-Pitch Gearing* (AGMA 370.01), American Gear Manufacturers Association, Alexandria, VA (1973).

3. *Gear and Spline Design*, IBM Standard Publication DEP 1-6000 BKT, International Business Machines Corp., Armonk, NY (1970).

8

Assembly and Operation

The integrity of an assembly rests not only on the proper design of the gear teeth but also on the gear meshing conditions. High quality gears may be manufactured and still fail in service. Failure analysis indications can point to problems relating to backlash, contact conditions, thermal expansion, or moisture absorption. A combination of these problems are also a possibility if consideration is not given each item in the initial design.

It is extremely important that all plastics gearing applications be tested before final acceptance of the design. The many factors involved can produce unexpected reactions, but all testing should be conducted in the exact form of its intended usage, including the lubricant, if used. Lubrication cannot be an afterthought but should be planned in advance. Accelerated testing can also alter reactions that may not be indicative of the final product. For example, the test speed chosen to reach full life quickly may generate excessive heat and contribute to early failure in the testing.

Backlash

A simple definition of backlash is the amount by which the width of a tooth space exceeds the thickness of the engaging tooth on the operating pitch circles. This is true for a particular tooth and space width; however, it is not a sufficient gage for the accuracy of a gear mesh. A more explanatory definition states that backlash is the play between mating tooth surfaces at the point of tightest mesh in a direction normal to the tooth surface when the gears are mounted in their specified positions. The implication is that the tightest mesh position must be determined.

With gears in position, several measurements are usually required to locate the tight mesh point. In ratios higher than 1:1, it may be a difficult but necessary exercise. One means is to mesh each gear during inspection to find and mark the area with the highest total composite variation or thickest tooth. Mating gears are matched in the assembly, the gear mark with the pinion mark. Backlash is then set at this position and the gear locations fixed. Any such scheme becomes more difficult as the gear ratio increases above 1:1.

Measurement of backlash is usually accomplished by fixing the shaft of one gear while rotating the shaft of the other gear in both directions. Suitable measurement fixtures allow the checking at each tooth position to determine the point of tightest mesh and the variation for comparison with the allowable backlash. The purpose of backlash is to prevent contact of teeth on both sides of the mating tooth space and subsequent rise in temperature due to jamming, adverse operating conditions, and early tooth wear. As discussed earlier, backlash is governed by the operating center distance and tooth thickness. If both mating gears have standard tooth thickness and are mounted on standard center distance, theoretical line-to-line contact occurs as well as the undesirable double flank tooth contact. Zero backlash theoretically exists in this condition.

Two methods are generally used to obtain proper backlash. The first is to thin the gear teeth when the standard center distance system is used. If the pinion has low numbers of teeth, only the gear teeth are thinned. Thus the pinion teeth will not be weakened by thinning.

The second method is to separate the mounting shafts an appropriate amount to allow for all possible adverse conditions. This effectively enlarges the standard center distance. Gears made to standard dimensions are usually used in this case.

It is important to provide backlash at every gear mesh. Determination of the amount of backlash should be made when the mesh is in the static position, when it can be easily measured. Enough should be provided so adequate backlash is present under operating conditions in which it will vary with speeds, load variations, mounting bearing runout, and especially with plastics gears, temperature increase.

Figure 8.1 shows two gears in tight mesh with double flank tooth contact. Figure 8.2 shows backlash in the shaded portion along the tooth flank.

AGMA presents two methods of calculating backlash depending on the situation facing the designer (1). The first method determines the minimum and maximum assembled backlash between teeth of two mating spur gears considering standard center distance and tolerance, tooth thickness, testing radius, and their tolerances. Note that its

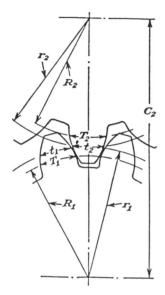

FIGURE 8.1 Representation of two gears in tight mesh. (From Ref. 2.)

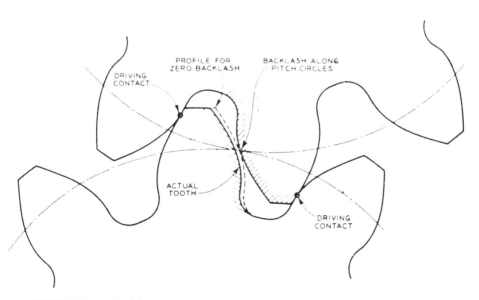

FIGURE 8.2 Backlash between two profiles. (From Ref. 1.)

TABLE 8.1 Minimum and Maximum Backlash

Minimum backlash = 2 tan ϕ ($\Delta R_P + \Delta R_G$ - tol C)

Maximum backlash = 2 tan ϕ ($\Delta R_P + \Delta R_G$ + tol C + tol R_{tP} + tol R_{tG})

where

ϕ = standard pressure angle
P = diametral pitch
ψ = helix angle
C = center distance
R_G = standard pitch radius of gear
R_P = standard pitch radius of pinion
R_{tP} = testing radius of pinion
R_{tG} = testing radius of gear
$\Delta R_P = R_P - R_{tP}$ = basic reduction from pitch radius
$\Delta R_G = R_G - R_{tG}$ = basic reduction from pitch radius
tol C = bilateral tolerance on center distance
P_n = normal diametral pitch
tol R_{tP} = tolerance on R_{tP}
tol R_{tG} = tolerance on R_{tG}
N_P, N_G = number of pinion, gear teeth respectively

External spur gears $C = \dfrac{N_P + N_G}{2P}$

Internal spur gears $C = \dfrac{N_G - N_P}{2P}$

Helical gears $C = \dfrac{N_P + N_G}{2P_n \cos \psi}$

Source: Ref. 1.

calculation determines only the maximum and minimum backlash values based on size for a particular design. This backlash range is the amount that would be expected for a particular static condition. Table 8.1 provides equations; Tables 8.2 and 8.3 provide example calculations; Tables 5.8, 7.6, 7.9, 7.10, and 8.4 provide necessary data. With this data, the procedures are relatively straightforward.

TABLE 8.2 Example of External Spur Gear – 20° Pinion Calculation

Item #	Specification	Source	Table Value or Calc.	Data Shown on Drawing
(1)	AGMA Quality Number	Given		Q8
(2)	Number of Teeth	Given		40
(3)	Diametral Pitch	Given		32
(4)	Pressure Angle	Given		20°
(5)	Standard Pitch Diameter	(2) ÷ (3)	40 ÷ 32	1.2500
(6)	Tooth Form	Given	"AGMA Std. Full Depth Involute."	
(7a)	Std. Cir. Thickness on Std. Pitch Cir.	Table 5.8	.0491	
(7b)	Reduction in Cir. Thickness (based on C.D. tolerance ± .0010)	Table 8.4	.00073	
(7c)	Max. Cir. Thickness on Std. Pitch Cir.	(7a) – (7b)		.04837
(8a)	Std. Pitch Radius	1/2 x (5)	.6250	
(8b)	Reduction from Pitch Radius (based on Cir. Thickness Reduction 7b)	Table 8.4	.0010	
(8c)	Testing Radius	(8a) – (8b)		.624
(8d)	Tolerance on Testing Radius[a]	Table 7.9		+.000-.0029
(9)	Total Composite Tolerance	Table 7.10		.0023
(10)	Tooth to Tooth Composite Tolerance	Table 5.8		.0010
(11a)	St. Addendum for (3)	Table 5.8	.0313	
(11b)	Outside Diameter	2 x [(8c) + (11a)]		1.311
(11c)	Tolerance on Outside Diameter	Table 7.6		+.000-.006

$N_P = 40$ $N_G = 120$

$P = 32$ Ratio = 3:1

$$C = \frac{N_P + N_G}{2P} = \frac{40 + 120}{2(32)} = 2.500 \text{ center}$$

AGMA quality number = Q8   Distance = $2.500 \pm .001$ (center distance tolerance from Table 7.15)

[a]In no case should the testing radius tolerance be less than the total composite tolerance. Testing radius tolerance is considered in these examples to be 1.25 times the total composite tolerance.

Source: Ref. 1.

TABLE 8.3 Example of External Spur Gear — 20° Gear Calculation

Item #	Specification	Source	Table Value or Calc.	Data Shown on Drawing
(1)	AGMA Quality Number	Given		Q8
(2)	Number of Teeth	Given		120
(3)	Diametral Pitch	Given		32
(4)	Pressure Angle	Given		20°
(5)	Standard Pitch Diameter	(2) ÷ (3)	120 ÷ 32	3.7500
(6)	Tooth Form	Given	"AGMA Std. Full Depth Involute."	
(7a)	Std. Cir. Thickness on Std. Pitch Cir.	Table 5.8	.0491	
(7b)	Reduction in Cir. Thickness (based on C.D. tolerance ±.0010)	Table 8.4	.00073	
(7c)	Max. Cir. Thickness on Std. Pitch Cir.	(7a) - (7b)		.04837
(8a)	Std. Pitch Radius	1/2 x (5)	1.8750	
(8b)	Reduction from Pitch Radius (based on Cir. Thickness Reduction 7b)	Table 8.4	.0010	
(8c)	Testing Radius	(8a) - (8b)		1.8740
(8d)	Tolerance on Testing Radius[a]			+.000-.0034
(9)	Total Composite Tolerance	Table 7.9		.0027
(10)	Tooth to Tooth Composite Tolerance	Table 7.10		.0011
(11a)	Std. Addendum for (3)	Table 5.8		.0313
(11b)	Outside Diameter	2 x [(8c)+(11a)]		3.811
(11c)	Tolerance on Outside Diameter	Table 7.6		+.000-.006

Minimum backlash = 2 tan ϕ (ΔRp + RG - tol C)

$\quad$ = .7279404 (.0010 + .0010 - .0010)

$\quad$ = .7279404 (.0010) = .00073

Maximum backlash = 2 tan ϕ (ΔRp + ΔRG + tol C + tol R_{tP} + tol R_{tG})

$\quad$ = .7279404 (.0010 + .0010 + .0010 + .0029 + .0034)

$\quad$ = .7279404 (.0093) = .0068

[a] In no case should the testing radius tolerance be less than the total composite tolerance. Testing radius tolerance is considered in these examples to be 1.25 times the total composite tolerance.

Source: Ref. 1.

TABLE 8.4 Basic Size Reduction, Fine-Pitch Spur Gears, 20° Pressure Angle

AGMA quality number	Reduce standard circular tooth thickness	Reduction from pitch radius
Q5	0.0018	0.0025
Q6	0.0015	0.0020
Q7	0.0011	0.0015
Q8	0.00073	0.0010
Q9	0.00066	0.0009
Q10	0.00058	0.0008
Q11	0.00051	0.0007
Q12	0.00044	0.0006
Q13	0.00035	0.0005
Q14	0.00029	0.0004
Q15	0.00022	0.0003
Q16	0.00015	0.0002

The second method calculates backlash on a given center distance and gives the actual backlash value in a known and measured system. If theoretical values are used in the equations, additional backlash will undoubtedly occur due to actual tooth thickness variations. See Table 8.5 for equations and Table 8.6 for examples. If computer design and analysis programs are available, techniques often are used to predict extremes of backlash conditions by inputting the worst conditions of tolerance and variation.

Probably the chief concern of the design engineer is to now know which backlash values to use in specific designs. The variations for concern are total composite tolerance for both pinion and gear, expansion due to temperature rise, moisture absorption, and runout of the mounting bearings. The approach is to start with the tight mesh

TABLE 8.5 Backlash and Center-Distance Relationships (center distance at which a pair of spur gears will engage with no backlash)

Case 1. Any two gears (standard or nonstandard tooth thicknesses)

$$\text{INV } \phi_1 = \text{INV } \phi + \frac{N_P (t + T) - \pi d}{d (N_P + N_G)} \qquad (1)$$

$$C_1 = \frac{\cos \phi}{\cos \phi_1} C \qquad (2)$$

$$C = \frac{N_P + N_G}{2P_d} \qquad (3)$$

Case 2. Two identical gears

$$\text{INV } \phi_1 = \text{INV } \phi + \frac{t}{d} - \frac{\pi}{2N_P} \qquad (1a)$$

$$C_1 = \frac{\cos \phi}{\cos \phi_1} C \qquad (2a)$$

ϕ = pressure angle, degrees
ϕ_1 = operating pressure angle, degrees
N_P, N_G = number of pinion and gear teeth respectively
t, T = tooth thickness of pinion and gear teeth respectively
d = pitch diameter of pinion
P_d = diametral pitch
C = standard center distance
C_1 = operating center distance

Source: Ref. 1.

center distance of a set of gears and adjust the center distance accordingly. Refer to Figures 8.3 through 8.5 and imagine the shaft of the gear being moved away from the shaft of the pinion. Where Figure 8.3 is the tight mesh condition, Figure 8.4 illustrates the amount of shaft separation by ΔC. Figure 8.5 illustrates the concept mathematically described in Table 8.1 This extended center distance has a direct bearing on the amount of backlash.

Total Composite Tolerances

When calculating the amount of backlash to allow for total composite tolerances, use the sum of one-half the total composite tolerance for each mating gear. The total composite tolerance factor is given as $(TCT_1/2) + (TCT_2/2)$. This factor assures that the minimum operating center distance is larger than the calculated tight mesh center distance as allowance for the gear inaccuracies.

A 77-tooth pinion drives a 120-tooth gear. They are AGMA Q7, 48 diametral pitch, and 20° pressure angle. Their standard pitch diameters are 1.604 and 2.5 inches. From Table 7.9 we find that respective total composite tolerances are 0.00293 and 0.00311 inches. One-half the sum of these values is 0.0030 inch. These data will be used later to illustrate further backlash procedures.

Thermal Expansion

The thermal expansion factor for the mating gear teeth should be calculated and its value also used to increase the tight mesh center distance. A procedure to determine the linear thermal expansion factor per degree temperature rise is found in Reference (4) as

$$C \left[(T - 70) \left(\frac{COEF_1 \text{ X } N_1}{N_1 + N_2} + \frac{COEF_2 \text{ X } N_2}{N_1 + N_2} - COEF_H \right) \right] \quad (8.1)$$

C = close mesh center distance
T = maximum temperature to which gears will be subjected (°F)
$COEF_1$ = coefficient of linear thermal expansion of material of 1st gear (in./in./°F)
$COEF_2$ = coefficient of linear thermal expansion of material of 2nd gear (in./in./°F)
$COEF_H$ = coefficient of linear thermal expansion of material of housing (in./in./°F)
N_1 = number of teeth in 1st gear
N_2 = number of teeth in 2nd gear

Coefficients of linear thermal expansion found in the materials properties charts and that of the housing are used in the equation. Note that the equation calls for values in degrees F whereas the charts usually give values in degrees C. To convert to the Fahrenheit scale from Celsius, multiply 9/5 by the temperature C and add 32° to the result. The calculated thermal factor is multiplied by the temperature increase from the dry, as-molded condition usually considered to be 70°F, to the maximum temperature rise. Assume that the previous gears have a maximum temperature rise in the application to

TABLE 8.6 Example of Center-Distance and Backlash Calculations

Given: 10 tooth pinion and 30 tooth gear, 100 diametral pitch, 20°
pressure angle, circular thickness of 0.01873 and 0.015708,
respectively.

Find: calculated center distance C_1 for no backlash.

$$INV\phi_1 = 0.014904 + \frac{10(0.01873 + 0.015708) - 3.1416(0.100)}{0.100(10 + 30)} \tag{1}$$

$$= 0.022459$$

$$\phi_1 = 22°49'$$

$$\cos\phi_1 = 0.92180 \qquad\qquad C = \frac{10 + 30}{2(100)} = 0.200$$

$$C_1 = \frac{(0.93969)}{(0.92180)}\ 0.200 = 0.2039 \text{ in.} \tag{2}$$

*Backlash Obtained Engaging Any Two Spur Gears on a Given Center
Distance*

$$\cos\phi_1 = \frac{C}{C_1}\cos\phi \qquad\qquad C = \frac{N_P + N_G}{2P_d} \tag{3}$$

$$\Delta INV\phi = INV\phi_1 - INV\phi$$

$$B_1 = 2C_1\left[\Delta INV\phi - \frac{C_1 - C}{C}\tan\phi\right]\quad \text{or}$$

$$B_1 = 2C_1\left[\Delta INV\phi + \frac{n/P_d - (t + T)}{2C}\right] \tag{4}$$

Example

Find: backlash

Given: Two 10 tooth pinions
50 diametral pitch — 20° pressure angle
0.03746 circular tooth thickness
0.2166 center distance

$$C = \frac{10 + 10}{2(50)} = 0.200$$

(continued)

TABLE 8.6 (continued)

$$\cos\phi_1 = \frac{(0.200)}{(0.2166)} \ 0.93969 = 0.86767 \tag{3}$$

$\phi_1 = 29.81°$

$INV\phi_1 = 0.052655$

$\Delta INV\phi = 0.052655 - 0.014904 = 0.037751$

$$B_1 = 2(0.2166) \left[0.037751 - \frac{(0.2166 - 0.2000)}{0.2000} \ 0.36397 \right] \tag{4}$$

$\quad = 0.0033$ in.

Source: Ref. 1.

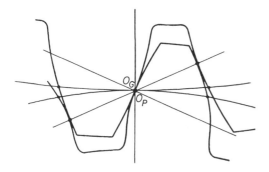

FIGURE 8.3 Pitch line contact. (From Ref. 3.)

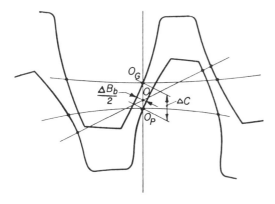

FIGURE 8.4 Increase in center distance. (From Ref. 3.)

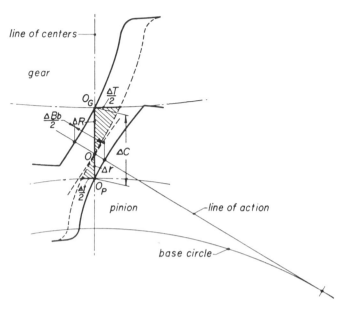

FIGURE 8.5   Involute triangles for increase in center-distance. (From Ref. 3.)

100°F; materials are a commonly used combination of nylon pinion and acetal gear. The coefficients of linear thermal expansion are for

Nylon: 5.0×10^{-5} in./in./degrees F
Acetal: 4.5×10^{-5} in./in./degrees F
Steel housing: 6.5×10^{-6} in./in./degrees F and the bearing T.I.R. is 0.0005

We now can calculate the thermal factor using our previous example data.

$$\text{Coef}_1 N_1/N_1 + N_2 = (5.0 \times 10^{-5})77/77 + 120$$

$$= 1.95 \times 10^{-5}$$

$$\text{Coef}_2 N_2/N_1 + N_2 = (4.5 \times 10^{-5})120/77 + 120$$

$$= 2.74 \times 10^{-5}$$

The gear coefficient factors are added and then reduced by the amount of the housing coefficient of linear thermal expansion. In this example there is a thermal expansion factor of 4.04×10^{-5} in./in./°F and for the 30° temperature rise a total of 0.001214 inch.

Moisture Absorption

In cases in which the plastics gear material is susceptible to expansion due to moisture absorption from high humidity conditions or other moisture exposure situations, use

$$\left(\frac{M_1 \times N_1}{N_1 + N_2} + \frac{M_2 \times N_2}{N_1 + N_2} \right) \tag{8.2}$$

N_1 = number of teeth in first gear
N_2 = number of teeth in second gear
M_1 = expansion due to moisture pick-up of material of first gear (in./in.)
M_2 = expansion due to moisture pick-up of material of second gear (in./in.)

Examination of materials property charts show water absorption as a percentage of weight gain for a 0.125-inch thick sample immersed in water for 24 hours. The sample is dried at 50°C and the weight compared to the dry weight. This rate is meaningless to the gear engineer for use with the moisture absorption equation given above. If there is a chance of high humidity conditions, the material supplier should be consulted to obtain an acceptable expansion value in inch/inch.

Table 8.7 is a listing of materials ranked according to water absorption rates, starting with the lowest values up to the higher rates for the common gearing materials and some of their modified conditions. These values are questionable for use in gear calculations, and materials selection decisions should be made with caution; however, the ranking shows the wide disparity of water absorption rates. Note that for various conditions of acetal, the rates are fairly constant, whereas for nylons and their various conditions, the water absorption rates can range from a low of 0.07% up to 1.9%.

Assume values of expansion due to moisture absorption of 2.5 X 10^{-4} in./in. for nylon and 5.0 X 10^{-5} in./in. for acetal in our previous example, and the calculations are

$$\frac{M_1 \times N_1}{N_1 + N_2} + \frac{M_2 \times N_2}{N_1 + N_2}$$

$$[(2.5 \times 10^{-4} \times 77) \div (77 + 120)] + [(5.0 \times 10^{-5} \times 120) \div (77 + 120)] =$$

$$9.77 \times 10^{-5} + 3.045 \times 10^{-4} = 1.28 \times 10^{-4} \text{ in./in.}$$

This value is also multiplied by the 30° temperature rise to give a total expansion due to moisture absorption of 3.84 X 10^{-3}.

TABLE 8.7 Material Ranking by Water Absorption Rate

Material	Rate
Polytetrafluoroethylene	0.0
Polyethylene: medium density	<0.01
high density	<0.01
high molecular weight	<0.01
low density	<0.015
Polyphenylene sulfides (40% glass filled)	0.01
Polyester: thermosetting and alkyds	
low shrink	0.01-0.25
glass — preformed chopping roving	0.01-1.0
Polyester: linear aromatic	0.02
Polyphenylene sulfide: unfilled	0.02
Polyester: thermoplastic (18% glass)	0.02-0.07
Polyurethane: cast liquid methane	0.02-1.5
Polyester synthetic: fiber filled — alkyd	0.05-0.20
glass filled — alkyd	0.05-0.25
mineral filled — alkyd	0.05-0.50
glass-woven cloth	0.05-0.50
glass premix chopped	0.06-0.28
Nylon 12 (30% glass)	0.07
Polycarbonate (10-40% glass)	0.07-0.20
Styrene-acrylonitrile copolymer (20-33% glass filled)	0.08-0.22
Polyester thermoplastic:	0.09
thermoplastic PTMT (20% asbestos)	0.10
glass sheet molding	0.10-0.15
Polycarbonate <10% glass	0.12
Phenolic cast: mineral filled	0.12-0.36
Polyester alkyd: asbestos filled	0.14
Polycarbonate: unfilled	0.15-0.18
Polyester cast: rigid	0.15-0.60
Acetal: TFE	0.20

(continued)

TABLE 8.7 (continued)

Material	Rate
Nylon 6/12 (30-35% glass)	0.20
6/10 (30-35% glass)	0.20
Polyester alkyd vinyl ester thermoset	0.20
Styrene-acrylonitrile copolymer: unfilled	0.20-0.30
Polycarbonate ABS alloy	0.20-0.35
Phenolic cast: unfilled	0.20-0.40
Acetal copolymer	0.22
homopolymer	0.25
Nylon 12 (unmodified)	0.25
Acetal (20% glass)	0.25-0.29
Poly(ancide-imide)	0.28
Acetal (25% glass)	0.29
Nylon 11 (unmodified)	0.30
Polyester elastomer	0.30-0.60
Polyimide	0.32
Nylon 6/12 (unmodified)	0.40
6/10 (unmodified)	0.40
Polyester-thermosetting and alkyds (cast flexible)	0.50-2.5
Nylon 6 (cast)	0.60-1.2
Polyurethane elastomer thermoplastic	0.70-0.90
Nylon 6/6: MOS_2	0.80-1.1
30-35% glass	0.90
unmodified	1.1-1.5
nucleated	1.1-1.5
Nylon 6 (30-35% glass)	1.3
unmodified	1.3-1.9
nucleated	1.3-1.9
Nylon 6/6-6 (copolymer)	1.5-2.0

Center Distance

Any discussion of center distance begins with the standard center distance and should include the operating center distance and tight mesh center distance. Because of the close relationship of tooth thickness, backlash, and tolerance on the center distance, various aspects have been covered in other sections. Theoretical center distance of the standard center distance system can be found using

$$C = N_p + N_g / 2\, P_d \qquad\qquad (8.3)$$

using values from our example,

$$C = 77 + 120/2 \times 48 = 2.0520833$$

This is equivalent to a tight mesh center distance if tooth thickness is the same on both gears. Tight mesh center distance is of interest during inspection as discussed in relation to the running gage type of inspection in Chapter 10.

On the other hand, operating center distance is the actual dimension between the centers of two shafts on which mating spur gears are mounted. The point of contact then is determined by tooth thickness and actual tooth contact. Equations in Table 8.1 and Table 8.5 can be used in calculations regarding physical relationships.

ΔC Calculation

The various factors discussed can be combined as follows (4).

$$\Delta_c = \frac{TCT_1 + TCT_2}{2} + C\left[(T-70)\left(\frac{COEF_1 \times N_1}{N_1 + N_2} + \frac{COEF_2 \times N_2}{N_1 + N_2} - COEF_H\right) + \right.$$

$$\left.\left(\frac{M_1 \times N_1}{N_1 + N_2} + \frac{M_2 \times N_2}{N_1 + N_2}\right)\right] + TIR \qquad\qquad (8.4)$$

where

Δ_c = required increase in center distance
TCT_1 = maximum total composite tolerance of first gear
TCT_2 = maximum total composite tolerance of second gear
C = close mesh center distance
T = maximum temperature to which gears will be subjected ($^\circ$F)
$COEF_1$ = coefficient of linear thermal expansion of material of first gear (in./in./$^\circ$F)

$COEF_2$ = coefficient of linear thermal expansion of material in second gear (in./in./°F)

$COEF_H$ = coefficient of linear thermal expansion of material of housing (in./in./°F)

N_1 = number of teeth in first gear

N_2 = number of teeth in second gear

M_1 = expansion due to moisture pick-up of material of first gear (in./in.)

M_2 = expansion due to moisture pick-up of material of second gear (in./in.)

TIR = maximum allowable runout of bearings

Since the tight mesh center distance is adjusted by the thermal expansion factor and the moisture absorption expansion factor, the bearing runout and total composite tolerance factors are added to ensure operation at worst case.

Example Calculation

In our example, calculations are

Total composite tolerance = 0.0030
Temperature rise = 30°F
Bearing runout = 0.0005
Thermal expansion factor = 4.04×10^{-5} in./in./°F
Expansion due to moisture absorption = 1.28×10^{-4} in./in.
Tight mesh center distance = 2.0520833 in.

With these values ΔC calculates to 0.0139 inch using equation (8.4). Shafts are then separated by this amount.

From the tight mesh center distance, we now have an operating center distance of $C + \Delta C$.

$$C' = C + \Delta C = 2.052083 + 0.0139 = 2.065983$$

$$\phi = 20 \text{ degrees}$$

$$\text{Cos } \phi' = C \cos \phi / C' = 2.0520833(.9396926)/2.065983 = 0.9333703$$

$$\phi' = 21.0336° \quad \text{INV } \phi' = 0.0174345$$

We have assumed in this example that the drive and driven shaft can be moved apart as described. The operating center distance (C') can now be used in the following equations to finalize part design. Using gear ratio (m_g) and operating center distance, the gear size is

determined. No tooth thinning will be used but balancing of tooth
thickness is possible to keep root stress at the same level for both
gear and pinion.

$$M_g = N/n \tag{8.5}$$

$$d' = \frac{2 C'}{m_g + 1} \tag{8.6}$$

$$D' = \frac{2 C' m_g}{m_g + 1} \tag{8.7}$$

$$r' = d'/2 \tag{8.8}$$

$$R' = D'/2 \tag{8.9}$$

$$\phi' = 21.0336°$$

$$INV\phi' = INV\phi_1 \text{ in equations (8.13) and (8.14)}$$

$$= 0.0174345$$

Data for standard gears provide radius of profile, tooth thickness
at that radius, and involute at the pressure angle.

$$D_1 = N/P = 120/48 = 2.5 \tag{8.10}$$

$$R_1 = D_1/2 = 1.25 \tag{8.11}$$

$$T_1 = \pi/2P = 0.0327249 \tag{8.12}$$

$$\phi = 20°$$

Operating tooth thickness of pinion and gear are calculated using

$$t' = 2r'[(T_1/2R_1) + INV\phi_1 - INV\phi] \tag{8.13}$$

$$T' = 2R'[(T_1/2R_1) + INV\phi_1 - INV\phi] \tag{8.14}$$

Where backlash has been previously calculated as in this example,
tooth thickness can be adjusted to balance tooth strength, as shown
in Figure 5.24 and

$$INV\phi_1 = \frac{n(t' + T' + B) - \pi d'}{(n + N)d'} + INV\phi \tag{8.15}$$

where

   C' = operating center distance
   d' = operating diameter of pinion

D' = operating diameter of gear
m_g = gear ratio
r' = operating radius of pinion
R' = operating radius of gear
R_1 = given radius of profile
t' = operating tooth thickness of pinion
T' = operating tooth thickness of gear
T_1 = given tooth thickness at R_1
n = number of pinion teeth
N = number of gear teeth
P = diametral pitch
B = backlash
ϕ = pressure angle at R_1
ϕ_1 = operating pressure angle
$INV \phi_1$ = involute of pressure angle at R
$INV \phi'$ = involute of pressure angle at operating radius

Contact Ratio

Contact ratio is a numerical index of the existence and degree of continuity of action (5). A minimum contact ratio of 1.2 will ensure smooth operation of the gear set. Since at least one pair of gear teeth is in contact at all times, load or motion is transferred without acceleration and deceleration.

Not only is contact ratio an indication of smoothness of action, it also determines the degree of load sharing. The maximum load that any one set of teeth must carry occurs when that pair of teeth is engaged at the pitch circle. When more than one set of gear teeth are in mesh, load sharing occurs. This is desirable, especially in plastics gearing in situations involving few numbers of teeth and large load-carrying requirements.

Tooth bending due to overloading is thought to bring adjoining teeth into contact sooner than with rigid metallic gearing. Proper design allowing for required power transmission is preferred over designs relying on tooth load sharing to distribute the adverse effects of tooth inaccuracies and gross manufacturing disparities. The magnitude of gear inaccuracy can lead to failure due to severe bending, abnormal root stress levels, and accelerated wear due to unplanned and abnormal sliding of contact surfaces of the teeth in mesh.

Table 8.8 can be used to quickly determine contact ratio for 20° full-depth nonundercut spur gears. The table value can be selected for each gear in the mesh and the total is the contact ratio (5).

To calculate contact ratio, the following equations can be used.

$$R_{b_1} = R_1 \cos \phi_1 \qquad\qquad (8.16)$$

$$R_{b_2} = R_2 \cos \phi_2 \tag{8.17}$$

$$p_{b_1} = 2\pi R_{b_1}/n \tag{8.18}$$

$$p_{b_2} = 2\pi R_{b_2}/n \tag{8.19}$$

$$m_p = \left(\sqrt{R_{o_1}^2 - R_{b_1}^2} + \sqrt{R_{o_2}^2 - R_{b_2}^2} - C \sin \phi_1\right)\Big/p_{b_1} \tag{8.20}$$

where

R_{b_1}, R_{b_2} = base radius of pinion and gear

R_1, R_2 = pitch radius of pinion and gear

ϕ_1 = operating pressure angle

p_{b_1}, p_{b_2} = base pitch of pinion and gear

n, N = numbers of teeth

R_{o_1}, R_{o_2} = outside radius of pinion and gear

C_1 = operating center distance

In the previous example, contact ratio calculates to be 1.794. Using Table 8.8 contact ratio is 1.844, only a 3% difference which is entirely acceptable since the values are well above the recommended 1.2.

Angular Error

In a gear train, the point of interest is often the final shaft. The method presented in the AGMA *Fine-Pitch Design Manual* is illustrated and described for determining angular error, sometimes referred to as lost motion (1). As described earlier, backlash values are to include effects of center distance tolerance, bearing clearance, gear tooth size, and total composite variations. Figure 8.6 shows pertinent data for our discussion. As Table 8.9 is developed, the symbol "S" (speed) refers to the rotational speed of any one shaft, relative to any other specific shaft. Since the usual practice is to assign the slowest shaft 1 speed, the output shaft "W" in Figure 8.6 is assigned 1 speed.
    Angular error θ is

$$\theta = Sc \sum \frac{B}{1/2\ D\ X\ S} + \frac{B'}{1/2\ D'\ X\ S'} +$$

$$\frac{B''}{1/2\ D''\ X\ S''} + \cdots + \cdots \tag{8.21}$$

TABLE 8.8 Contact Ratio For One 20-Degree Full-Depth Nonundercut Gear[a]

No. of teeth	Contact ratio	No. of teeth	Contact ratio	No. of teeth	Contact ratio	No. of teeth	Contact ratio	No. of teeth	Contact ratio	No. of teeth	Contact ratio
10	0.685	20	0.778	30	0.827	40	0.857	60	0.802	80	0.913
11	0.698	21	0.785	31	0.830	42	0.861	62	0.895	85	0.917
12	0.710	22	0.790	32	0.834	44	0.866	64	0.897	90	0.920
13	0.721	23	0.796	33	0.837	46	0.870	66	0.900	95	0.923
14	0.731	24	0.801	34	0.840	48	0.874	68	0.902	100	0.926
15	0.741	25	0.806	35	0.843	50	0.877	70	0.904	120	0.934
16	0.749	26	0.810	36	0.846	52	0.881	72	0.906	140	0.943
17	0.757	27	0.815	37	0.849	54	0.884	74	0.908	160	0.948
18	0.765	28	0.819	38	0.852	56	0.887	76	0.909	180	0.952
19	0.772	29	0.823	39	0.854	58	0.890	78	0.911	200	0.956

[a]Based on a standard addendum of a = 1/Pd. To get contact ratio of the mesh, add contact ratio of pinion to contact ratio of gear.
Source: Ref. 7.

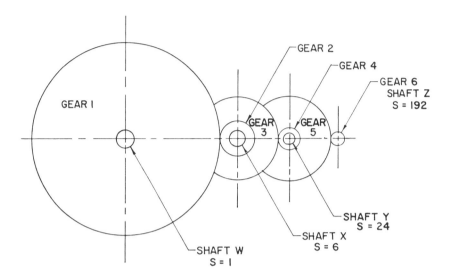

FIGURE 8.6 Gear train angular error.

Gear number	No. of teeth	Diametral pitch	Pitch diameter
1	144	48	3.0"
2	24	48	0.5"
3	80	64	1.25"
4	20	64	0.3125"
5	120	96	1.25"
6	15	96	0.15625"

(From Ref. 1.)

where:

B, B', B", etc. = backlash in inches at the pitch line between G and its mating gear

D, D', D", etc. = pitch diameters of the successive gears in the train in inches

S, S', S", etc. = speed of the shafts on which the gears D, D" are mounted

c = constant of proportionality

To get θ in radian, c = 1.0

To get θ in degrees, c = 57.296

To get θ in minutes, c = 3438

TABLE 8.9 Angular Error Calculation Example

(a)

Gear number	Pitch diameter (D)	Backlash of mesh (B)	Gear ratio	Shaft speed (S)	B / D/2XS	
1	3.0	0.0015	6:1	1S	0.0010	
3	1.25	0.0030	4:1	6S	0.0008	θ = 6.88 min
5	1.25	0.0030	8:1	24S	0.0002	θ' = 22°
					TOTAL	
					0.002	

(b)

Gear number	Pitch diameter (D)	Backlash of mesh (B)	Gear ratio	Shaft speed (S)	B / D/2XS	
1	3.0	0.0015	6:1	1S	0.0010	
3	1.25	0.0030	4:1	6S	0.0008	θ = 6.53 min
5	1.25	0.0015	8:1	24S	0.0001	θ' = 20.9°
					TOTAL	
					0.0019	

(c)

Gear number	Pitch diameter (D)	Backlash of mesh (B)	Gear ratio	Shaft speed (S)	B / D/2XS	
1	3.0	0.00075	6:1	1S	0.0005	
3	1.25	0.0030	4:1	6S	0.0008	θ = 5.16 min
5	1.25	0.0030	8:1	24S	0.0002	θ' = 16.5°
					TOTAL	
					0.0015	

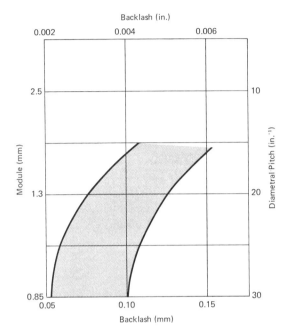

FIGURE 8.7 Suggested backlash for nylon gears. (From Ref. 6.)

From gear tooth ratios and speed designations we find that shaft X will turn at 144/24 X 1 = 6 speed; shaft Y at 80/20 X 144/24 X 1 = 24 speed; shaft Z at 120/15 X 80/20 X 144/24 X 1 = 192 speed. Therefore, θ = 1 X 3438 X 0.0020 = 6.88 minutes.

If shaft Z is held fixed, shaft W can be turned through an angle of 6.88 minutes due to backlash. If shaft W is held fixed, shaft Z can be turned through an angle of θ' if: θ' = 192 X 57.296 X 0.0020 = 22°. This can be seen in Table 8.9(a).

Table 8.9(b) shows that by reducing the backlash at mesh 5-6 by 50%, θ and θ' are reduced by 5%. If, however, the backlash at mesh 1-2 is reduced by 50%, θ and θ' are reduced by 25%, as shown in Table 8.9(c). Using this method, one can manipulate values and make corrections to provide the necessary angular accuracy for the application.

Recommended Backlash Values

Figure 8.7 is provided here as backlash values suggested for nylon gears if calculations are not made to determine the exact values needed.

TABLE 8.10 Recommended Minimum Normal Backlash for Straight
Bevel Gears

Diametral pitch (P_d)	Backlash (in.) AGMA quality number		Metric module (m)	Backlash (mm) AGMA quality number	
	4-9	10-14		4-9	10-14
1.00-1.25	0.032	0.024	25.0-20.0	0.80	0.60
1.25-1.50	0.027	0.020	20.0-16.0	0.64	0.48
1.50-2.00	0.020	0.015	16.0-12.0	0.48	0.36
2.00-2.50	0.016	0.012	12.0-10.0	0.40	0.30
2.50-3.00	0.013	0.010	10.0-8.00	0.32	0.24
3.00-4.00	0.010	0.008	8.00-6.00	0.24	0.18
4.00-5.00	0.008	0.006	6.00-5.00	0.20	0.15
5.00-6.00	0.006	0.005	5.00-4.00	0.16	0.12
6.00-8.00	0.005	0.004	4.00-3.00	0.12	0.09
8.00-10.0	0.004	0.003	3.00-2.50	0.10	0.08
10.0-12.0	0.003	0.002	2.50-2.00	0.08	0.06
12.0-16.0	0.003	0.002	2.00-1.50	0.06	0.05
16.0-20.0	0.002	0.001	1.50-1.25	0.05	0.04
20.0 and above	0.002	0.001	1.25 and below	0.04	0.03

Source: Ref. 7.

Further, the application should not involve critical timing, high speeds,
or exacting transfer of motion.

Backlash calculations for bevel gears involve an extremely complex
situation. Therefore Table 8.10 is given as a guide to recommended
minimum normal backlash based on average conditions for general purpose gearing and is measured at the outer cone distance on the gear
member. In bevel gears, backlash is directly affected by runout,
tooth accuracy, and by their axial position in the inspection machine
or the assembly.

References

1. *Design Manual for Fine-Pitch Gearing* (AGMA 370.01), American Gear Manufacturers Association, Alexandria, VA (1973).

2. E. Buckingham, *Analytical Mechanics of Gears*, Dover Publications, New York (1963).

3. A. H. Candee, *Introduction to the Kinematic Geometry of Gear Teeth*, Chilton Company, New York (1961).

4. W. McKinlay and S. D. Pierson, *Plastics Gearing*, ABA/PGT Publishing Company, Manchester, CT (1976).

5. D. W. Dudley, Ed., *Gear Handbook*, McGraw-Hill, New York (1962).

6. *DuPont Zytel Nylon Resin Design Handbook*, Bulletin E-44971, E. I. DuPont DeNemours & Co., Wilmington, DE.

7. *AGMA Standard System for Straight Bevel Gears* (AGMA 208.03), American Gear Manufacturers Association, Alexandria VA (1978).

9
Stock Gear Selection

Gearing may be available for purchase from supply houses as off-the-shelf items. Several companies publish and distribute catalogs listing plastic gears that can be purchased by catalog number. In some cases only a minimum amount of information is provided and it is left to the purchaser to apply the part to the application properly.

Also, some of the typical gearing texts and publications deal only with very general and basic gear equations. The information here gives some guidance in what to expect in using standard off-the-shelf gearing.

Limitations and constraints to gear design may include the nature and location of mountings or shafts. Such specific questions must be answered as,

Are shaft positions fixed prior to design of the gears?
Will, or can, an idler gear be used?
What are the general magnitude of loads?
Can stock plastic gears even be considered due to environment?

Some mechanical systems dictate the location of gear mounting shafts that will necessitate manipulation of design procedures, possible relaxing of requirements, or compensation by using alternate approaches in the gear application. Since catalog gearing consists of parts generally made of the most common gear materials, standard pitches or modules, and recommended face widths, designs can often be limited.

Using a simple fine-pitch gear design example and preferred diametral pitches and modules, Table 9.1 shows the problem associated

TABLE 9.1 Gear Size Estimation

	British units $(N_P + N_G = 2 \times P \times C;$ center distance, $C = 2.375$ in.)				Metric units $(N_P + N_G = 2 \times C/m;$ center distance, $C = 60.3$ mm)		
Case no.	Preferred diametral pitch (P)	Pinion teeth (N_P)	Gear teeth (N_G)	Case no.	Preferred module (m)	Pinion teeth (N_P)	Gear teeth (N_G)
	Ratio = 1:1				Ratio = 1:1		
1	20	47.5	47.5	10	1.25	48.2	48.2
2	24	57.0	57.0	11	1.0	60.3	60.3
3	32	76.0	76.0	12	0.8	75.4	75.4
4	40	95.0	95.0	13	0.6	100.5	100.5
5	48	114.0	114.0	14	0.5	120.6	120.6
6	64	152.0	152.0	15	0.4	150.7	150.7
7	80	190.0	190.0	16	0.3	201.0	201.0
8	96	228.0	228.0	17	0.25	241.2	241.2
9	120	285.0	285.0				
	Ratio = 2:1				Ratio = 2:1		
18	20	31.7	63.3	27	1.25	32.2	64.3
19	24	38.0	76.0	28	1.0	40.2	80.4
20	32	50.7	101.3	29	0.8	50.3	100.5
21	40	63.3	126.7	30	0.6	67.0	134.0
22	48	76.0	152.0	31	0.5	80.4	160.8
23	64	101.3	202.7	32	0.4	100.5	201.0
24	80	126.7	253.3	33	0.3	134.0	268.0
25	96	152.0	304.0	34	0.25	160.8	321.6
26	120	190.0	380.0				
35	20	48.0	48.0	37	0.5	81.0	160.0
36	20	47.0	48.0	38	1.25	32.0	64.0

Case 35 — Ratio 1:1, C = 2.4 inches.
Case 36 — Ratio 1.02:1, C = 2.375 inches.
Case 37 — Ratio 1.98:1, C = 60.25 mm.
Case 38 — Ratio 2:1, C = 60 mm.

with fixed mounting shaft gearing. Center distance, diametral pitch, module, and ratio were arbitrarily selected for illustrative purposes only.

Study of Table 9.1 shows that there are possible designs for all but one diametral pitch combination with a 1:1 ratio, but only four with the 2:1 ratio requirement. The basic requirements are met if both the pinion and gear have full numbers of teeth. Likewise, for metric module (SI) gears, there is but one possible design at 1:1 ratio (Case 16) and only two at 2:1 ratio (Cases 30 and 33). Cases 35-38 illustrate how adjustments can be made to accommodate other preferred pitches or modules. In Case 35, the center distance was increased whereas in Case 36, a change in output ratio occurs. Similar changes from the initial requirements were made in Cases 37 and 38, but note that in Case 37, neither center distance nor ratio have been accommodated.

Changes as illustrated in Cases 35-38 are usually made in circumstances where another pitch is required and ratio or center distance requirements can be relaxed or compensated for in other parts of the design.

Actual gear designs are often quite restrictive in the same manner. The general procedure is to test designs until all the basic requirements are met. Then the estimated gear size is checked for adequate beam strength and wear life. This preliminary design exercise will show that there are usually only a few combinations of pitch or module that will meet all the design criteria. It is sometimes possible by certain manipulations to suit the design and use standard gears. This is done notably by using enlarged center distances, thinning of purchased gears, or accepting slight ratio changes that will be reflected in input/output relationships.

Computer Calculations

Most gearing calculations today are made quickly and efficiently using calculators or computers. Standard equations can provide somewhat restrictive designs. The recommended and preferred pitches and modules are number series, necessary for economical and practical reasons. In plastics gearing, where a mold must be produced for almost every application, standard restrictions for metal gears do not necessarily apply nor should they be considered mandatory, as seen in Table 9.2. In the generation of trial data done by computer, design and analysis programs need to have the same flexibility for decision making of alternate designs as would be available with design engineer decision making.

TABLE 9.2   Gear Size Estimation Using Nonstandard Pitches
and Modules

British units			Metric units			

British units: $(N_P + N_G = 2 \times P \times C;$ center distance, $C = 2.375$ in.)

Metric units: $(N_P + N_G = 2 \times C/m;$ center distance, $C = 60.3$ mm)

Case no.	Preferred diametral pitch (P)	Pinion teeth (N_P)	Gear teeth (N_G)	Case no.	Preferred module (m)	Pinion teeth (N_P)	Gear teeth (N_G)
	Ratio = 1:1				Ratio = 1:1		
1	20	47.5	47.5	4	0.9	67.0	67.0
2	19.368	46.0	46.0	5	0.6	100.5	100.5
3	20.211	48.0	48.0	6	0.603	100.0	100.0
				7	0.597	101.0	101.0
	Ratio = 2:1				Ratio = 2:1		
8	20	31.7	63.3	11	0.4	100.5	201.0
9	19.579	31.0	62.0	12	0.402	100.0	200.0
10	20.211	32.0	64.0	13	0.398	101.0	202.0

Case 4 — Nonpreferred module.

Catalog Selection

When it is determined that catalog gears can be used in the design, a study of available data from several companies should be made. Plastics gears selected from catalogs are made to standard dimensions and the mechanism center distance must be extended to compensate for the gear size. This practice is reasonable since the gear supplier cannot stock tooth sizes for every combination possible.

Quality class and gear size should be known but may not be provided by the gear supplier. These can be quickly and accurately determined using an inspection running-gage as described in Chapter 10. If tooth thickness at the standard pitch circle is in agreement with published values, the gears are considered to be "standard."

It is absolutely essential that an attempt is not made to use the
gears on the standard center distance if it is determined that they
are made to standard size. This would place the gears in a theoretical tight-mesh condition. Such a position forces the gears together
if they are at the standard size and contributes to problems such as
overheating, wear, system overload, and locking particularly if
loads are light. Some methods used where teeth are at the maximum
thickness are to increase the shaft center distance, slightly thin the
teeth of one or both of the gears, match the thickest tooth with the
widest space, or provide some mechanical device that will provide allowance during operation.

It should never be assumed that since the material is plastic, the
teeth will flex enough to provide proper tooth action. Such conditions indicate a questionable design practice that usually leads to
early failure within the system. Line-to-line contact may not permit
tooth deflection sufficient for rotation. If the forcing load is suffi-
ciently high, it will cause deflection in mountings, bearings, and
shafts. Double profile contact and jamming of the teeth into the
mating tooth space root diameter leads to early failure by wear.

Increased Center Distance vs. Tooth Thinning

Increasing the center distance or thinning teeth are the most common
solutions used to handle the problem of standard mating gear teeth.
We have already briefly discussed movement of the shaft centers.
This is referred to as the nonstandard-center-distance system and
is common practice when using stock gears. There may be situations
where this system is not feasible. For example, other elements in the
assembly may be fixed in design and it would be more costly or impractical to increase the center distance. It may also be more costly
for manufacturing when machining housings if bilateral tolerances
could not be applied. The "Standard Center Distance System" is
usually the preferred system. The center distance equals one-half
the sum of mating gear *theoretical* pitch diameters. This allows fixed
standard center distances with bilateral tolerances to be applied to
the gear shaft mounting holes of an assembly. Gear tooth thickness
is then reduced an amount sufficient to assure free running conditions when mounted on the low limit center distance. This reduced
tooth thickness is controlled by a specification of testing radius
slightly smaller than the standard pitch radius. Standardized center
distance tolerances and corresponding gear tooth reductions are studied in other chapters.

Not only gear tooth reductions are used in the standard center distance system but gear tooth increases are also possible. The ease of

producing gears or plastic gear molds with enlarged or reduced tooth
thicknesses often makes gear modification almost inconsequential as
far as cost is concerned. In cutting of metal gears or gear mold electrodes, it is usually a matter of making deeper or shallower cuts in
the gear or electrode blank. Standard cutting methods and machines
are used but modified cutters are used to cut the electrode blank.
The gears produced can both be reduced in tooth thickness, only one
reduced or enlarged, or a combination of reductions and enlargement
used for mating gears.

  Modifications must be such that gear replacement can be accomplished without difficulty if it becomes necessary some time in the
future. Modifications by reduction or enlargement of teeth are not
only economical in plastic gearing, but tooth thickness can be dupli-
cated for subsequent gear replacements from the same mold.

Molding vs. Cutting of Plastics Gears

There are many plastic gear applications that cannot be satisfied by
purchase of stock plastic gears. Design requirements may be such
that special modifications or materials require molding or cutting of
plastic gears. Where gears of enormous size are in constant contact
with corrosive elements and failures mean long down-time for repairs,
plastic gears have been molded in sections and bolted together, or
large cut plastic gears have been used successfully (1,2). In fine-
pitch gearing, cutting can present a whole new set of problems. Because of the small size, the gear blank may move, tool tip radii may
wear excessively, or the cutting tool will cause feathering on the tooth
edges. The user may find accelerated wear with cut gears over molded
gears. Some users have tested products using molded gears only to
experience a high field failure rate. Production gears were cut rather
than molded. In relation to Nylatron®, Polymer Corporation states
that "data on the two methods of manufacture is not interchangeable"
between injection-molded gearing and cut tooth gearing. If there is
doubt as to the choice between cutting and molding as a manufacturing method, consultation with a reliable plastics processor is necessary. Since various machine capacities differ and part configuration
influences the mold design, the number of cavities and gear size must
be considered. The manufacturer can provide valuable assistance as
to the various economies and improvements to increase wear life or
part strength. Further elaboration can be found in reference (3) in
the introduction to a discussion of machining and cutting.

  Cutting has a number of disadvantages. Some materials, when cut,
tend to feather on the edges. Often the teeth will deflect during cutting if they are small and have a narrow face width, thus causing
tooth shape inaccuracies. Usually plastic gears must be cut at a

slower speed than metallic gears to avoid heat buildup and eliminate grooving of the teeth during hobbing. Many plastic gears running against steel pinions experience accelerated wear rates when the outer surface is removed by hobbing or shaping. This "skin" effect covers and protects the inner material. Tool marks provide minute stress raisers and can expose any voids or molding imperfections leading to rapid failure. Also, tools often become dull quite quickly if the gear material is glass-filled since they expose glass fibers and tear out glass beads. In fine-pitch gears, the small tool points can be quickly worn in cutting plastic gears. On the other hand, small gears in large quantities can be produced with molding. Molding sizes vary from the finest pitches to enormous gear sectors that are cut and bolted together to form power gears. The geometry of each gear can be designed to give significant cost savings since intricate shapes can be molded in combination with other mechanical parts. Cam surfaces, mounting shafts without bearings, pawls, stops, splines, and levers are all possible. In most cases the part will be produced so that additional machining will not be necessary.

Contrary to thoughts against the use of cut plastic gears, there is much in the literature concerning technological advances that have made plastic gear cutting a routine operation (4,5,6). Most notably is the publication of tool dimensions and feeds and speeds for cutting equipment. A significant publication that covers gear design only for cut-tooth Nylatron gearing should be consulted if large, cut gears are being considered (7). Because recommendations vary based on the selected material, careful consideration should be given to minimize any cutting difficulties. Specifically, data for hobbing and shaping will be of interest in tooth machining, where some recommendations are to use data similar to machining of bronze gears.

Because of the economy of molding gears in one operation, the use of metal inserts is discouraged. There are applications, however, that benefit by their use. Heat dissipation in high-speed gearing, close bore-to-shaft fits, stability of the gear on the shaft, improved load carrying capacity, and controlled shrinkage are all valuable uses of the metal insert. When considering the use of an insert, the designer must be assured that the advantages outweigh the added cost of the insert and the cost to insert it in the mold during the molding operation. For the molding of gears, part configuration and tooth shrinkage guidelines must be followed.

Subjects of Consideration

The following is a list of subjects generally related to plastics gear selection and application that can be used as a check list to insure that they are not overlooked in the design.

1. Environment — This includes not only the lubricant and the general atmosphere in which the gears will operate, but also the possibility of debris entering the system, that may be either abrasive or detrimental to the lubricant or plastic. This contamination may also be in the form of a liquid or gas.

2. Stress from undercut has been discussed elsewhere, but if gears of low numbers of teeth are purchased, caution is advised. Molding can induce stresses between the undercut portion of the tooth and the thicker portion near the pitch line. The stress level at manufacture is unknown, and monitoring for crazing during testing is recommended.

3. The availability of off-the-shelf gearing is limited. Certain materials with fillers, lubricants, and combinations are not usually readily available.

4. Attachment systems are extremely limited, requiring cautious design if the usage of stock gearing is anticipated. The advantage of molding two or more parts into a single unit cannot be had. Any attachment schemes are limited to what the supplier has to offer.

5. Quality class of gears is usually limited. If this is not a concern to the particular application, the gears may be purchased and inspected to determine the quality class for analysis purposes.

6. Teeth are usually of the involute form. If modifications are necessary due to speed, loads, or deflections, stock gears normally will not suit the application.

7. Most stock gear companies are knowledgeable as to the necessity for mold tooth corrections. Buyers must assure themselves that this is the case and that the finished gears have the correct tooth form.

8. Tolerances should be known for critical design applications. Aging of the gears, or shelf life, may have stress relaxation effects that should be known. Although this is a subject of some controversy, it is noted here for the designer's consideration.

9. Thermal and moisture absorption factors must be considered as in all good gearing design.

Stock, or off-the-shelf, gearing serves a real need in the gearing industry, and the cautions above should not deter the gear engineer considering their usage. As in designing for molding or cutting of plastics gears, attention to the above details help to insure a reliable and safe application.

References

1. *Power Transmission Design*, Lamont Gear Company, De Pere, WI (1982).

2. *Lawton Power Transmission Products*, C. A. Lawton Company, Norristown, PA (December 1980).

3. *The Celcon Acetal Copolymer Design Manual*, Celanese Plastics Company, Newark, NJ (1969).

4. *DuPont Delrin Acetal Resin Design Handbook*, E. I. DuPont DeNemours and Co., Inc., Wilmington, DE (1981).

5. *DuPont Zytel Nylon Resin Design Handbook*, Bulletin E-44971, E. I. DuPont DeNemours and Co., Inc., Wilmington, DE.

6. G. W. Michalec, *Precision Gearing Theory and Practice*, John Wiley and Sons, New York (1966).

7. *Nylatron Nylon Gear Design and Fabrication Manual*, BR-36D, Polymer Corporation, Reading, PA (1983).

10

Inspection

Plastic gears are inspected, tested, and classified by the same methods as metal gears. Therefore use of the AGMA quality classification system is recommended. It provides a well-defined means of discussing gearing aspects without ambiguity or confusion of terms.

The tables in Chapter 7 provided inspection tolerances for qualities Q3 for coarse-pitch gears and Q5 through Q16 for fine-pitch gears. The higher the Q value, the smaller the tolerance values allowed. Quality class is an easy reference for relative tolerances.

Tooth-to-tooth and composite tolerance values should be specified on the drawing, especially if high-quality gearing is necessary in the application and if specific values were used for calculations and cannot be exceeded without jeopardizing the application.

Coarse-pitch gears may be physically measured by scaling features of the gear. Runout, pitch, tooth thickness, profile, index, and lead can be checked, but for each feature checked, a tolerance must be provided. It is also necessary to understand how each element of the gear and analytical checks affect the output from the gear mesh. Reference (1) discusses inspection methods and reference (2) discusses many of the element effects. If transmission of power is the only consideration, as is sometimes the case, sophisticated analysis may not be necessary. Element checking is done using the same measuring and gaging tools as for metal gears. Care must be used to make sure that the soft plastics surfaces are not flexed or indented and that incorrect readings recorded. Element analysis is also covered extensively in reference (1).

Composite Inspection

Gears in the fine-pitch range, that of 20 diametral pitch and finer, are usually measured on a variable-center-distance fixture. Called a composite check, the sample gear is run in intimate contact with a master gear of known accuracy. Radial displacement or variation in center distance is shown on a dial indicator or paper tape recording mechanism (see Figure 10.1). This method of measurement approximates the gear under actual operating conditions. Because it is running against a high-quality master gear, variations are considered to be produced by inaccuracies in the gear being inspected. This measuring method is called a composite check because the result is the combined effect of the following variations:

    Runout
    Tooth thickness
    Lateral runout (sometimes called wobble)
    Pitch
    Profile

Since the composite check measures only radial variations, it does not measure the effect of lead or tooth index variations.

During the composite check, tooth-to-tooth composite variation appears as "flicker" on the dial indicator of a variable-center-distance fixture as the gear is rotated through 360°. Flicker is the result of the following variations:

    Circular pitch
    Tooth thickness
    Profile

Total composite variation consists of runout, lateral runout, and tooth-to-tooth composite variation. In other words, this variation is the total center distance displacement read on the indicating device; i.e., the difference between the maximum and minimum reading in 360° of gear rotation. Also, the tooth-to-tooth composite value is the difference between the maximum and minimum reading for the tooth mesh with the largest difference. A paper chart recording device attached to the variable-center-distance fixture, which is related to a master gear of known accuracy during the inspection, removes most of the human element from the inspection process. Such charts are often provided with the purchase of master gears and with high-quality production gears. A sketch representing a recording chart is shown in Figure 10.2.

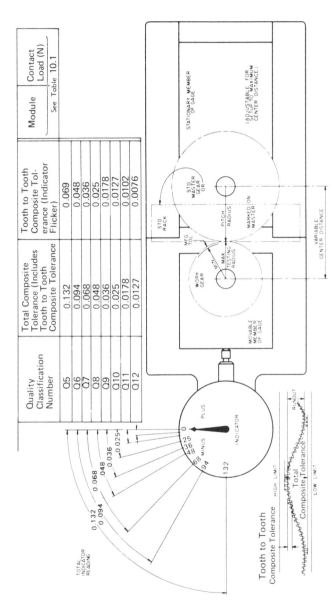

FIGURE 10.1 Variable-center-distance gage. Graphic record of the indicator reading for a gear showing how total composite tolerance can be analyzed to determine runout and tooth-to-tooth composite tolerance caused by circular pitch error, thickness variation and profile error. Gage center distance with indicator set at zero is set up to the sum of the pitch radius of the master gear and the maximum testing radius of gear as specified. When a master rack is used, the indicator is set at zero when the distance from the pitch-line of the master to the axis of the work gear equals the maximum testing radius. Note: Tolerance values are in millimeters. (From Ref. 3.)

Quality Classification Number	Total Composite Tolerance (Includes Tooth to Tooth Composite Tolerance	Tooth to Tooth Composite Tolerance (Indicator Flicker)	Module	Contact Load (N) See Table 10.1
Q5	0.132	0.069		
Q6	0.094	0.048		
Q7	0.068	0.036		
Q8	0.048	0.025		
Q9	0.036	0.0178		
Q10	0.025	0.0127		
Q11	0.0178	0.0102		
Q12	0.0127	0.0076		

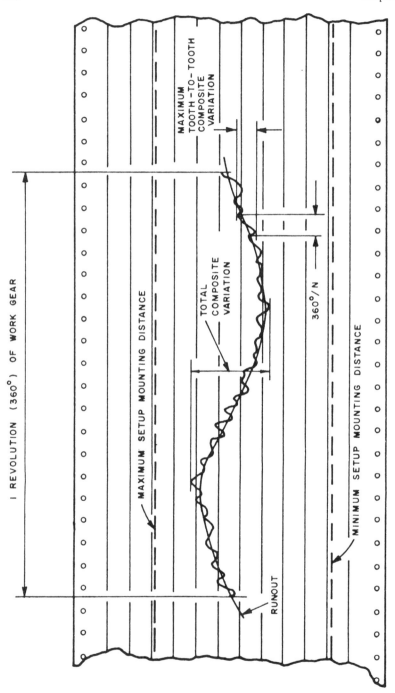

FIGURE 10.2 Typical chart of gear tooth variations. (From Ref. 1.)

TABLE 10.1 Recommended Contact Load for Inspection of
Fine-Pitch Spur Gears on Variable-Center-Distance Gage

Module	Contact load in newtons*	
	Steel	Plastics
2.50 to 1.26	8.9	4.4
1.25 to 0.76	7.8	3.9
0.75 to 0.61	6.7	3.4
0.60 to 0.51	5.6	2.8
0.50 to 0.41	4.4	2.2
0.40 to 0.31	3.3	1.6
0.30 to 0.20	2.2	1.1

Diametral pitch	Contact load in ounces*	
	Steel	Plastics
10 to 19	32	16
20 to 29	28	14
30 to 39	24	12
40 to 49	20	10
50 to 59	16	8
60 to 79	12	6
80 to 100	8	4

*Loads are for gears with a face width of 2.5 mm or greater
in the SI metric system and 1.0 inch or greater in the English
system. For widths less than 2.5 mm or 1.0 inch, the contact
load should be reduced proportionately or determined by actual test and agreed on by gear manufacturer and user.

Checking Loads

Variable-center-distance fixtures are designed so that the load applied between the master and gear being inspected is adjustable. Only
enough load is used during inspection to maintain intimate contact between the gears. Excessive load may deflect the teeth and mounting
shafts if they are small, as is the usual case with many fine-pitch
gears. Table 10.1 gives load in newtons for SI metric gears of 2.5

mm face width and gives load in ounces per 0.1-inch gear face width. Although these loads are generally recommended in the gear industry and by the AGMA, the value for each gear design should be agreed upon by the gear molder and the user. The table values listed for plastics are one-half those recommended for metal gears.

Testing Radius

An important calculation must be made for the set-up of the variable-center-distance gage. Since the gear can have increased or decreased tooth thickness on the standard pitch circle, a specification of testing radius must be determined. Testing radius added to the radius of the master gear then provides the set-up distance for the variable-center-distance gage.

Calculation of testing radius for spur gears is as follows. The values which must be known are:

1. Maximum tooth thickness on the pitch circle of the gear being inspected, t_G
2. Number of teeth in gear, N_g
3. Pressure angle of gear, ϕ
4. Maximum tooth thickness on the pitch circle of the master gear, t_m. $t_m = p/2$ for standard master; p = circular pitch of the master gear
5. Number of teeth in master gear, N_m

Master gears up to 48 diametral pitch usually have a 3.0-inch pitch diameter and those of 48 diametral pitch and finer have a 1.5-inch pitch diameter.

1. Calculate standard center distance, C_1.

$$C_1 = (N_G + N_m/2 P_d \text{ where } P_d \text{ is diametral pitch.} \tag{10.1}$$

2. Calculate operating pressure angle, ϕ_2 from

$$INV\phi_2 = \frac{P_d(t_G + t_m) - \pi}{N_g + N_m} + INV\phi \tag{10.2}$$

3. Calculate tight mesh center distance with the master gear, C_2.

$$C_2 = C_1 (\cos \phi / \cos \phi_2) \tag{10.3}$$

4. Calculate the maximum testing radius.

$$R_t \text{ (max)} = C_2 - (0.5 \text{ X } N_m/P_d) \tag{10.4}$$

Therefore the set-up center distance on a variable-center-distance gage is the sum of the pitch radius of the master gear and the testing radius of the gear being tested. Example calculations of a gear and pinion were shown in Figures 8.4 and 8.5.

The composite method of checking gears does not give individual values for the five variations mentioned above. Due to the small size of fine-pitch gears, analytical checking is specified only in extreme cases. Analytical checking is never to be specified in conjunction with the composite check because inspection values are not compatible and comparisons cannot be made between the two methods. References (1) and (4) provide detailed discussion and alternate procedures for analytical checking.

The designer should recognize that added inspection time, special equipment, data analysis, and therefore extra cost are involved in analytical checks. They should never be required if composite checks will be adequate.

Gear metrology itself is a branch of the gearing industry that cannot be covered completely in this text. Because the majority of plastics gears can be measured as discussed above, such subjects as single-flank testing, portable inspection devices, and undulation measurement are left to the user of special gears.

TABLE 10.2 Spur and Helical Master Gear Quality Classes Suggested for Use with AGMA Quality Numbers

Fine-pitch gearing		Coarse-pitch gearing	
Quality number	Master gear quality class	Quality number	Master gear quality class
5-8	1	6-9	1
9	2	10	2
10	3	11	3
11	4	12	4
12	5	13	5
13-16	6	14	6-7

Source: Ref. 1.

TABLE 10.3 Coarse-Pitch Spur and Helical Master Gears, Arbor Mounted Type

CLASSIFICATION BY COMPOSITE TOLERANCE								
Class		1	2	3	4	5	6	7
Total	Thru 4.25 P D	.00070 ref	.00050 ref	.00035 ref.	.00025	.00018	.00013	.00009
	Over 4.25 Thru 8.25 P D	.00085 ref.	.00060 ref.	.00042 ref.	.00030	.00022	.00016	.00011
Tooth-To-Tooth	Thru 4.25 P D	.00030 ref.	.00020 ref.	.00014 ref.	.00010	.00007	...	...
	Over 4.25 Thru 8.25 P D	.00040 ref.	.00025 ref.	.00020 ref.	.00012	.00009	...	...

TOLERANCE ON GEAR ELEMENTS							
Lead Tolerance[a] (based on comparison with a control master to the lead tolerance table)							
Tooth Length / Class	1	2	3	4	5	6	7
1"	.00023	.00020	.00018	.00017	.00016	.00015	.00014
2"	.00028	.00025	.00023	.00022	.00021	.00020	.00019
3"	.00033	.00030	.00028	.00026	.00025	.00025	.00024
4"	.00038	.00035	.00033	.00032	.00031	.00030	.00029

INVOLUTE PROFILE TOLERANCE									
(based on comparison to a control master)									
Class / Pitch	4	6	8	10	12	14	16	18	20
1	.00035 max.	.00035 max.	.00035 max.	.00035	.00032	.00029	.00027	.00025	.00023
2	.00035 max.	.00035 max.	.00030	.00026	.00024	.00022	.00020	.0002	.0002
3	.00031	.00027	.00024	.00021	.00020	.0002	.0002	.0002	.0002
4	.00026	.00022	.00020	.0002	.0002	.0002	.0002	.0002	.0002
5	.00022	.00020	.0002	.0002	.0002	.0002	.0002	.0002	.0002
6	.00020	.0002	.0002	.0002	.0002	.0002	.0002	.0002	.0002
7	...	...	...	...	...	...	...	...	...

				Tolerance	
				Pitch Diameter	
Gear Element	Tooth Thickness Grade	Pressure Angle		Thru 4.25	Over 4.25 Thru 8.25
Arc Tooth Thickness[b]	Grade A	All Pressure Angles		− .0002	− .0003
	Grade B	All Pressure Angles		± .0025	± .003
Runout (tiv) Over One Pin	Grades A & B	13° Thru 19° P A		.0004	.0005
		Over 19° P A		.0003	.0004
Pitch	Grades A & B	All Pressure Angles		.0001	.00015

TOLERANCE ON BLANK ELEMENTS							
Class	1	2	3	4	5	6	7
Bore[c] + Tolerance −.0000	.0001	.0001	.0001	.00005	.00003	.00003	.00003
Outside Radius	.0005	.0005	.0005	.0005	.0005	.00025	.00025
Face Runout Per In. of Radius .0001 Min.	.00015	.00015	.00015	.00015	.00005	.00005	.00005
OD Runout	.0002	.0002	.0002	.0002	.0001	.0001	.0001

[a]Lead: 0.0001 additional end easing allowed either end of face, 80% central face shall be within tolerance shown.
[b]Arc tooth thickness: After determination of arc tooth thickness, necessary adjustment to the OD shall be calculated from the equations shown for the calibration of master gears.
[c]Bore: The difference between the effective bore size and the size between any two diametrically opposite points shall not be more than the bore tolerance. Bell mouth will be allowed on 10% of the total bore length, with a length of bell mouth not to exceed 0.250 total.
Source: Ref. 1.

TABLE 10.4 Fine-Pitch Spur and Helical Master Gears, Arbor Mounted Type

CLASSIFICATION BY COMPOSITE TOLERANCE [8]						
Class	1	2	3	4	5	6
Tooth-To-Tooth	.0002 ref.	.00014 ref.	.00010 ref.	.00007	.00005	.00004
Total	.0005 ref.	.00035 ref.	.00025 ref.	.00018	.00013	.00009

TOLERANCE ON GEAR ELEMENTS						
Class	1	2	3	4	5	6
Pitch	.00012	.00010	.00008	1	1	1
Profile[2]	.00015	.00013	.00010	.00010	.00010	.00010
Runout (tiv)[3]	.00030	.00021	.00015	1	1	1
Bore[(4 6)]	.0001	.0001	.00005	.00003	.00003	.00003
Outside Rad.[4]	.0005	.0005	.0005	.0005	.00025	.00025

Gear Element	Tooth Thickness Grade	Pitch Diameter	Tolerance
Arc Tooth Thickness [7]	Grade A	Thru 2.25	−.0002
		Over 2.25	−.0003
	Grade B	±1.25% of Circular Pitch	
Lead — Max. Total 5	.0001		
Face Runout (tiv)	.00005 (.0001 Max.)		

1. Classification shall be by composite tolerance in this area.

2. Not applicable 50 DP and finer. The profile tolerances shown are predicated on comparison of the master-gear profile to the profile of a control master.

3. Over one pin.

4. Tolerance all plus.

5. Predicated on comparison with a control master, .0001 additional end easing is allowed at either end of face. 80% central face shall be as shown.

6. BORE: The difference between the effective bore size and the size between any two diametrically opposite points shall not be more than the bore tolerance. Bell mouth will be allowed on 10% of the total bore length with a length of bell mouth not to exceed .250 total.

7. ARC TOOTH THICKNESS: After determination of arc tooth thickness, necessary adjustment to the OD shall be calculated from the formulae shown for the calibration of master gears. See Appendix A, Master Gears, page A-6.

Value is based on Arc Tooth Thickness $= \dfrac{C_p}{2}$.

8. When composite check is specified for a master it becomes necessary to check out this master with another master of like or higher quality class. Unless it is clearly stated that the master must have a total-composite-tolerance level, classes 1 through 3 shall be considered acceptable on the basis of gear-tooth element checks only. Classes 4 through 7 can only be accepted on the basis of both composite check and the gear-tooth-element check as described above.

9. When considering master gears with fewer than 20 teeth, check with the manufacturer regarding calibration and use.

Source: Ref. 1.

Master Gears

When using the variable-center-distance fixture, the assumed perfect master gear becomes a major factor in the inspection. Tables 10.2, 10.3, and 10.4 provide the necessary master gear tolerances for each class of gear being inspected.

Element Inspection

Reference (1) covers inspection procedures for individual element checks and details of the composite action checks (Part III). Appendix

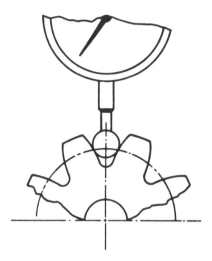

FIGURE 10.3 Single-probe runout check for spur and helical gears. (From Ref. 1.)

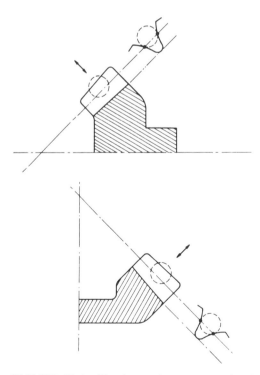

FIGURE 10.4 Single-probe runout check for bevel and hypoid gears. (From Ref. 1.)

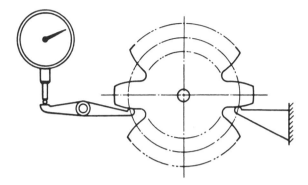

FIGURE 10.5   Two-probe check for spur and helical gears.   (From Ref. 1.).

A of the same reference provides supplementary data that may be needed in extreme cases and valuable information on index, pitch, and spacing relationships data that is often helpful in analysis of lost motion or angular error situations.

Runout can be measured by indicating over pins, balls, or other devices placed in successive tooth spaces.   Single- or double-probe gauges are also available and are illustrated in Figures 10.3 through 10.6 for spur, helical, bevel, and hypoid gears.   In the single-probe

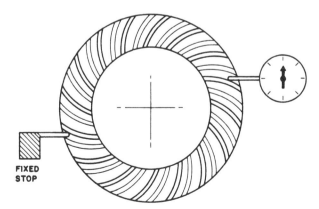

FIGURE 10.6   Two-prove check for bevel and hypoid gears.   (From Ref. 1.)

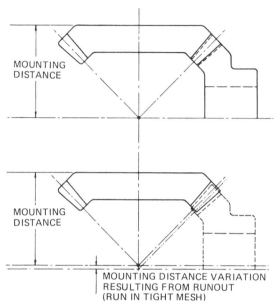

FIGURE 10.7 Variation in mounting distance, runout check of bevel or hypoid pinion. (From Ref. 1.)

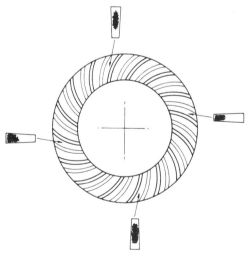

FIGURE 10.8 Runout contact pattern variation. Shifting of tooth contact shows presence of runout. Sound variation also characterizes the existence of runout. (From Ref. 1.)

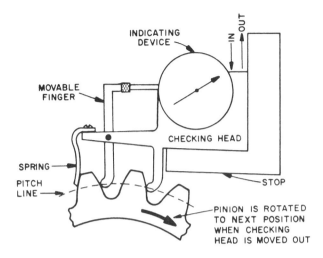

FIGURE 10.9 Pitch measuring instrument schematic. (From Ref. 1.)

method, the readings include some effects of eccentricity, out-of-roundness, axial runout, and profile, spacing, tooth thickness, and spiral or helix angle variations. In the double-probe method, the difference between the highest and lowest readings represent twice the amount of runout when making the check in one complete revolution. Bevel and hypoid gears are often checked for runout using the rolling check, whereby the center distance or mounting distance variation is observed between the gear being inspected and a gear of known accuracy. The gears are rolled together in tight mesh with one member loaded on a movable center. Runout is sometimes measured by indicating the root circle or outside diameter when these surfaces have been machined simultaneously with the tooth profiles. A contact pattern check may be used to observe a variation in contact pattern by running the gears in a testing machine. The tooth contact will shift progressively around the gear from heel to toe and from toe to heel. A variation in sound occurs in every revolution (see Figures 10.7 and 10.8).

Pitch tolerance (tooth spacing and index) can be measured using dividing heads of suitable accuracy or pitch-checking instruments. Measurements are made at or near the pitch circle, preferably on the loaded side if the gear operates in only one direction (see Figure 10.9).

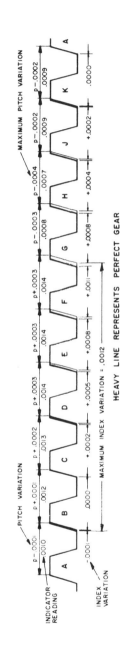

HEAVY LINE REPRESENTS PERFECT GEAR

TEETH	#1 INDICATOR READINGS	#2 SPACING VARIATION	#3 PITCH VARIATION	#4 INDEX VARIATION
A — B	.000	.0002	-.000	-.000
B — C	.0012	.0001	+.000	.0000
C — D	.0013	.0001	+.0002	+.0002
D — E	.0014	.0000	+.0003	+.0005
E — F	.0014	.0000	+.0003	+.0008
F — G	.0014	.0006 MAX	+.0003	+.0011
G — H	.0008	.0001	-.0003	+.0008
H — J	.0007	.0002	-.0004	+.0004
J — K	.0009	.0000	-.0002	+.0002
K — A	.0009	.0001	-.0002	.0000
10	.0010		MAX PITCH VARIATION	

CHECK. TOTAL = O   CHECK LAST NUMBER = O

.0011 = AVERAGE INDICATOR READINGS (FOR CORRECT PITCH "P")

(−) + .0001
.0012 = MAX INDEX VARIATION

1. Set indicator to .0010 on first pair of teeth and enter as first reading in Column No. 1.
2. Record successive tooth-to-tooth indicator readings around gear (Column No. 1.)
3. Spacing variation (Column No. 2) is the difference between successive indicator readings in Column No. 1.
4. Total all indicator readings and divide by number of teeth to find average spacing which represents the correct pitch.
5. Pitch Variation (Column No. 3) is the difference between each indicator reading of Column No. 1 and the average spacing.
6. To find Index Variation (Column No. 4) add successive Pitch Variation from Column No. 3.
7. To find maximum Index Variation, algebraically subtract the maximum and minimum values in Column No. 4.

FIGURE 10.10 Pitch and index measurement comparison, same gear. (From Ref. 1.)

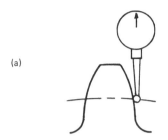

(a)

(b)

(c)

FIGURE 10.11 Measured profiles for index, pitch, and spacing variation. (a) Index (I) (angular position), (b) pitch (P) (angular position or average spacing (S)), (c) spacing variation (S_1) (tooth-to-
tooth spacing tolerance). (From Ref. 1.)

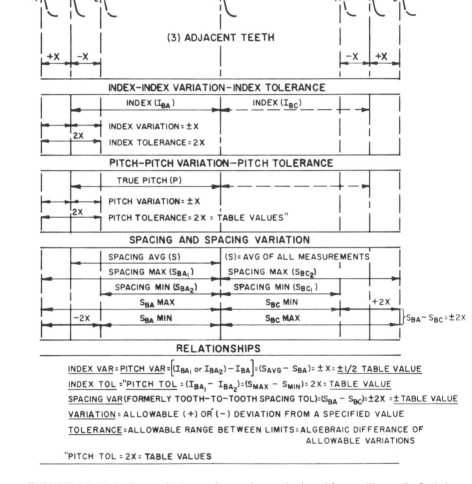

FIGURE 10.12 Index, pitch, and spacing relationships. (From Ref. 1.)

For index checking, angular-positioning devices such as index heads, dividing plates, optical polygons, and thoedolites are available to determine spacing variations. Figure 10.10 compares tabulated and

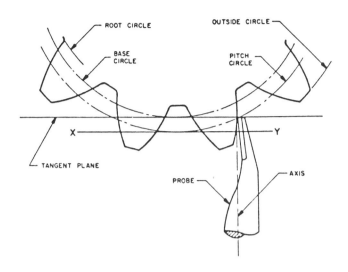

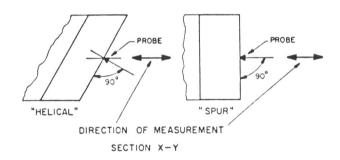

FIGURE 10.13 Involute profile measuring probe. (From Ref. 1.)

graphical pitch and index measurements and provides an interpreta-
tion of the data. It shows the method of calculating pitch from measured values of spacing or index. As additional clarification to pitch
tolerance determination, Figure 10.11 and Figure 10.12 are provided
and are self-explanatory.

Profile measurement instruments contact a tooth profile and provide
a chart representation of the actual profile as it deviates from the true
involute. Figure 10.13 shows the probe set at the pitch line. Typical

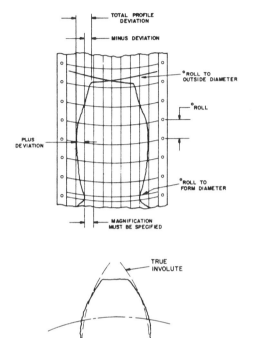

FIGURE 10.14   Tooth profile and corresponding chart.   (From Ref. 1.)

charts are shown in Figures 10.14 and 10.15. Profile measurements
are useful in determining tip modifications and for checking undercut
conditions.

Several alternate methods have been used to check spur and helical gear profiles. Figures 10.16 through 10.19 illustrate methods
whereby incremental measurements are taken and compared to com-
puted values. Figure 10.20 illustrates comparison of a scale tooth
layout to the gear tooth optically magnified and projected for visual
comparison, although this is usually limited to small gears.

Lead checking on spur, helical, and herringbone gears is accomplished by advancing a probe along a tooth surface, parallel to the
axis, while the gear rotates in a specified timed relationship, based
on the specified lead. Figure 10.21 is a representation showing lead,
lead angle, and helix angle relationships. Figures 10.22 through
10.31 are provided to indicate various charts that can be obtained

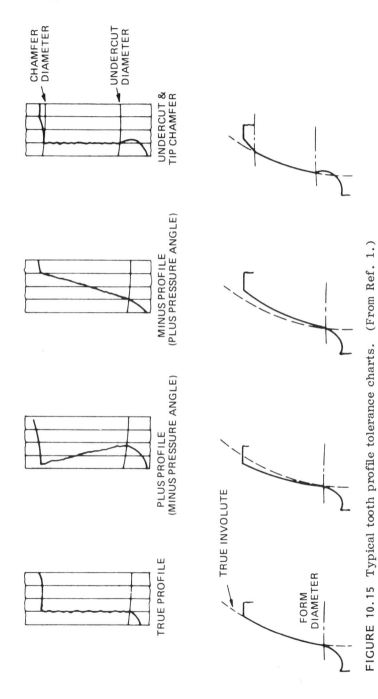

FIGURE 10.15 Typical tooth profile tolerance charts. (From Ref. 1.)

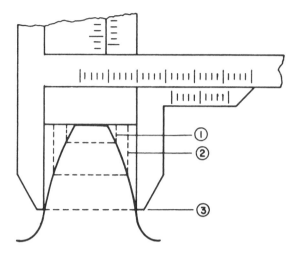

FIGURE 10.16 Tooth caliper measurement. (From Ref. 1.)

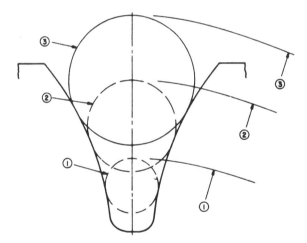

FIGURE 10.17 Measurement by pins or balls. (From Ref. 1.)

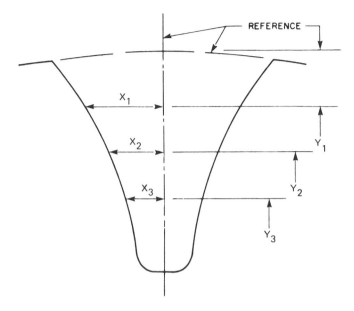

FIGURE 10.18 Coordinate measurement. (From Ref. 1.)

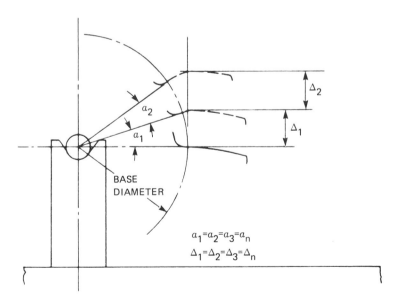

FIGURE 10.19 Line-of-action measurement. (From Ref. 1.)

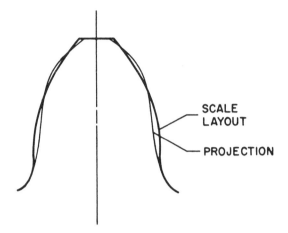

FIGURE 10.20 Gear tooth projection. (From Ref. 1.)

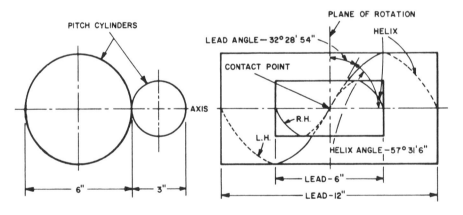

FIGURE 10.21 Lead, lead angle, and helix angle. (From Ref. 1.)

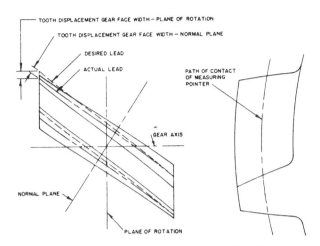

FIGURE 10.22 Tooth displacement of helical gear. (From Ref. 1.)

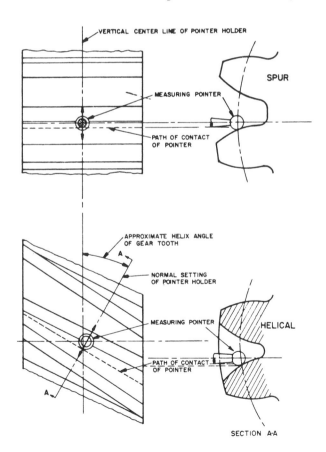

FIGURE 10.23 Lead probe check. (From Ref. 1.)

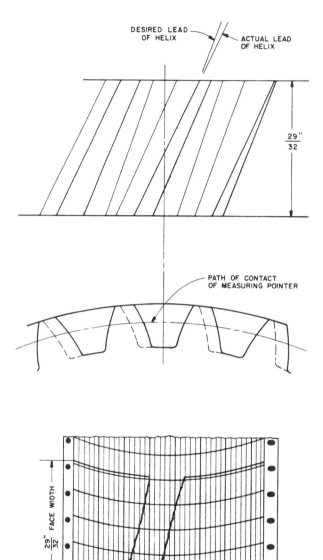

FIGURE 10.24 Right-hand helical gear teeth, short lead, (-). (From
Ref. 1.)

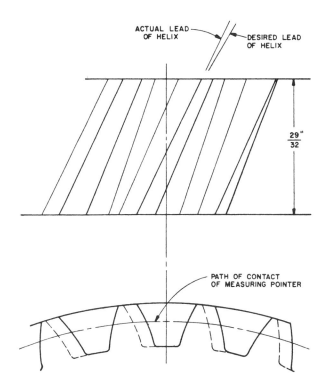

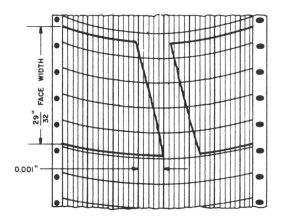

FIGURE 10.25 Right-hand helical gear teeth, long lead, (+). (From Ref. 1.)

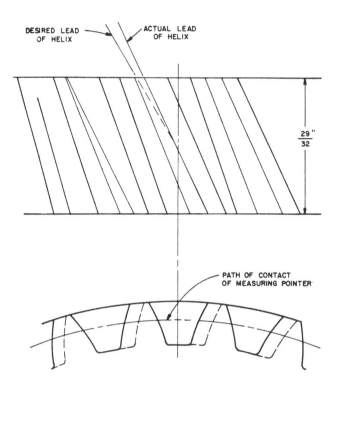

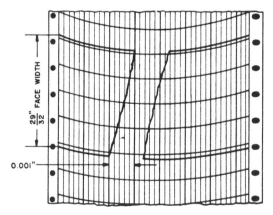

FIGURE 10.26 Left-hand helical gear teeth, long lead, (+). (From Ref. 1.)

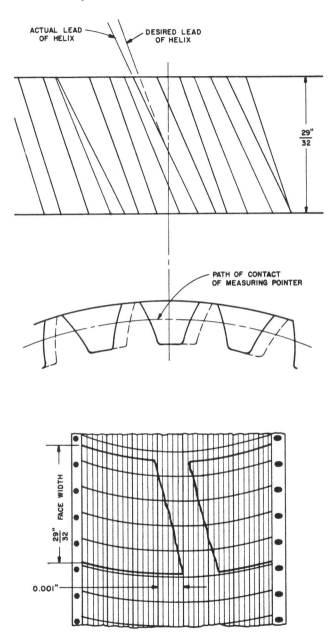

FIGURE 10.27 Left-hand helical gear teeth, short lead, (-). (From Ref. 1.)

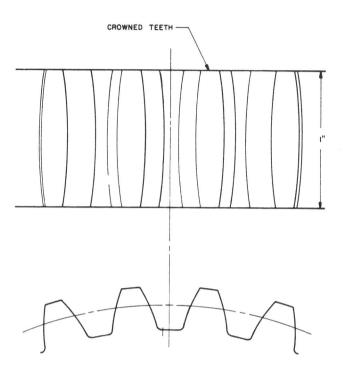

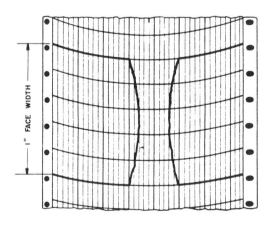

FIGURE 10.28 Spur gear, crowned teeth. (From Ref. 1.)

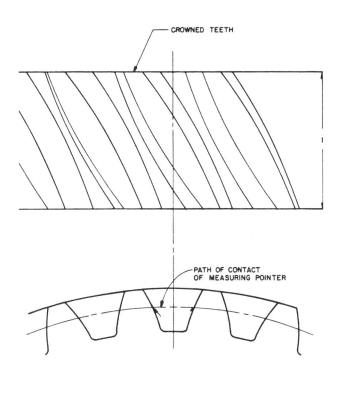

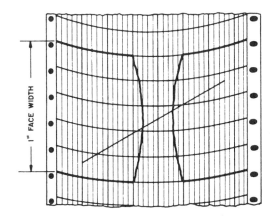

FIGURE 10.29 Helical gear, crowned teeth. (From Ref. 1.)

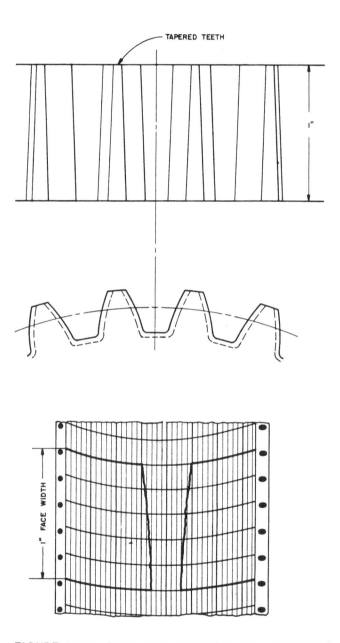

FIGURE 10.30 Spur gear, tapered teeth. (From Ref. 1.)

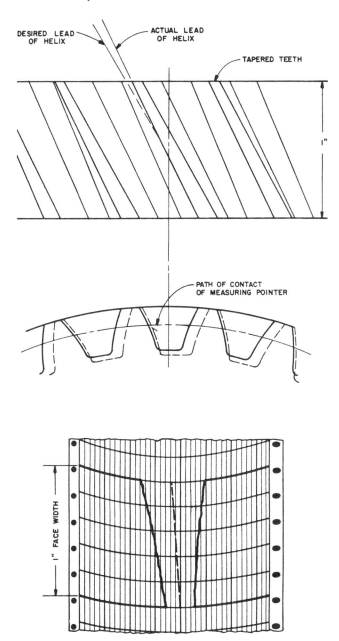

FIGURE 10.31 Helical gear, tapered teeth and lead deviation. (From Ref. 1.)

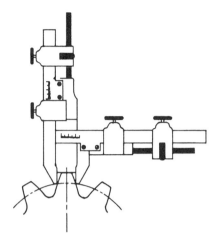

FIGURE 10.32 Gear caliper tooth thickness measurement. (From Ref. 1.)

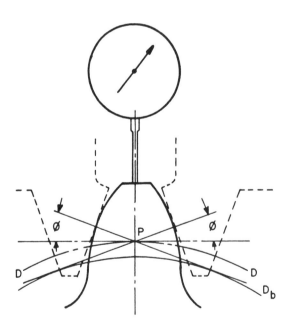

FIGURE 10.33 Gear tooth comparator tooth thickness measurement. (From Ref. 1.)

when plastics are used for helical, herringbone, and worm gears. Also, problems of taper are possible with spur gears with wide faces when cavities produce taper. An alternate method of lead checking is to coat a gear tooth with dye and run it against a master or mating gear. The resulting pattern reveals the amount of face contact and accuracy of the lead.

Tooth thickness can be measured using vernier gear-tooth calipers (Figure 10.32), addendum comparator (Figure 10.33), span measurement (Figure 10.34), or by measurement over pins (Figure 10.35). In both the tooth capiter and addendum comparator method, the outside diameter is used as a reference dimension and could interject some inaccuracy in the reading. Tooth measurement readings must be compared to calculated values. The overwire or pin measurement is an accurate method whereby measurements are not influenced by outside diameter variation or by runout of the outside diameter. Errors in tooth spacing and profile are accounted for in the readings.

Tooth contact pattern techniques are used to check the conditions of contact between mating gears. This is a common and recommended method for checking bevel gears that is extremely accurate. Figure 10.36 shows typical patterns and Figure 10.37 illustrates methods of determining horizontal or vertical movements to adjust the location of the pattern. If during the production of a gear, a contact pattern is obtained similar to one of the undesirable patterns of Figure 10.36, (c) through (p), adjustments in the gear machine will correct the pattern. In preparing molds for plastics gears, it is recommended that knowledgeable and experienced mold producers be consulted.

Overwire Measurement

Many publications provide table values for measurements over wires, pins, or balls for standard gears using selected sizes of wires, pins, or balls (d_w). For gears with an even number of teeth, the measurement is calculated by

$$M = \frac{D \cos \phi}{\cos \phi_1} + d_w \tag{10.5}$$

and gears with an odd number of teeth by

$$M = \frac{D \cos \phi}{\cos \phi_1} \cos \frac{90°}{N} + d_w \tag{10.6}$$

The value ϕ_1 is calculated by

$$INV\phi_1 = \frac{t}{D} + INV\phi + \frac{d_w}{D \cos \phi} - \frac{\pi}{N} \tag{10.7}$$

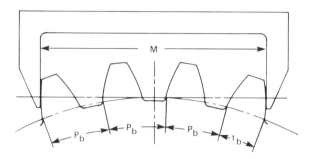

FIGURE 10.34 Span measurement of tooth thickness. (From Ref. 1.)

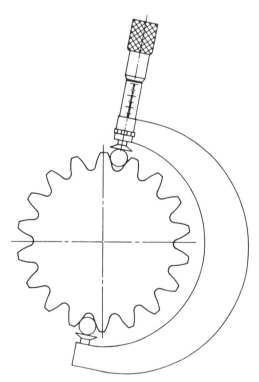

FIGURE 10.35 Tooth thickness by overwire measurement. (From Ref. 1.)

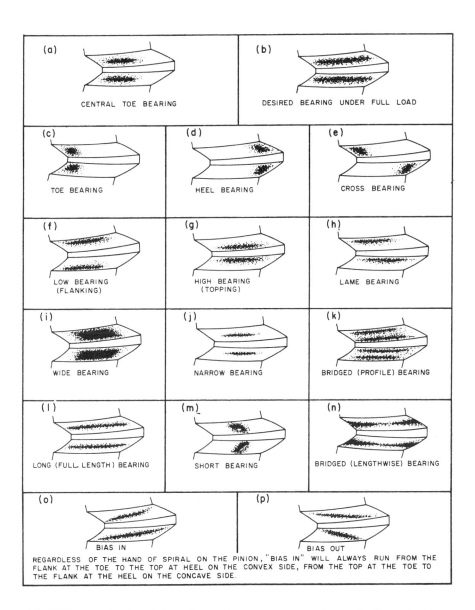

FIGURE 10.36 Tooth bearing patterns on the pinion teeth. A left-hand pinion is used throughout. The patterns are representative of those on a right-hand pinion or a straight bevel pinion. (From Ref. 1.)

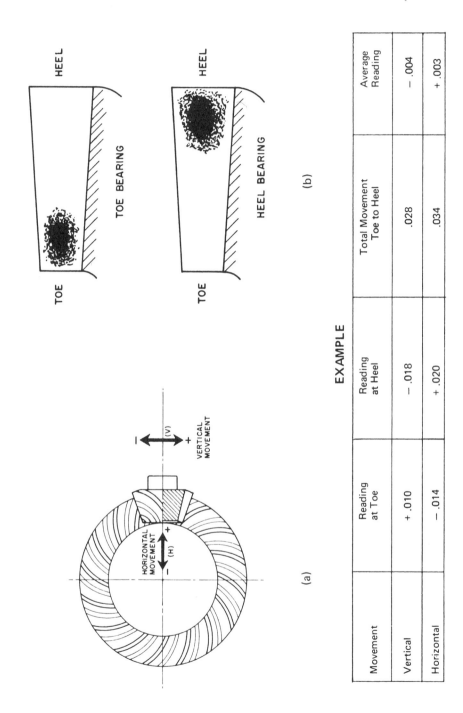

(a)

(b)

EXAMPLE

Movement	Reading at Toe	Reading at Heel	Total Movement Toe to Heel	Average Reading
Vertical	+ .010	– .018	.028	– .004
Horizontal	– .014	+ .020	.034	+ .003

FIGURE 10.37 Tooth contact pattern adjustment procedure. (a) Explanation of V and H movements, (b) nomenclature.

Total vertical movement = (+.010) - (-.018) = +.028

Total horizontal movement = (-.014) - (+.020) = -.034

The algebraic signs of these totals are ignored since the magnitude of these quantities is the item of interest.

$$\text{Average vertical setting} = \frac{(+.010) + (-.018)}{2} = -.004$$

$$\text{Average horizontal setting} = \frac{(-.014) + (+.020)}{2} = +.003$$

When obtaining the average readings, both the magnitude and the direction (sign) are important. The average reading is used to place the tooth bearing in the center of the tooth, and to observe its appearance, or to compare it with the average reading for a master set of gears for the same job. (From Ref. 1.)

If the measurement is to determine tooth thickness, the calculations are

$$t = D \left(\frac{\pi}{N} + INV\phi_c - INV\phi - \frac{d_w}{D \cos \phi} \right) \tag{10.8}$$

where

$$\cos \phi_c = \frac{D \cos \phi}{2 R_c} \tag{10.9}$$

For gears with an even number of teeth

$$R_c = \frac{M - d_w}{2} \tag{10.10}$$

For gears with an odd number of teeth

$$R_c = \frac{M - d_w}{2 \cos (90°/N)} \tag{10.11}$$

The term R_c is as shown in Figure 10.38 and is the dimension from the center of the part to the center of the measuring pin.

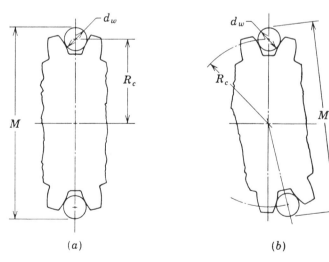

(a)   (b)

FIGURE 10.38 Geometry of overwire measurement: (a) even number of teeth, (b) odd number of teeth. (From Ref. 2.)

References

1. *AGMA Gear Handbook, Volume 1: Gear Classification, Materials, and Measuring Methods for Unassembled Gears* (AGMA 390.03), American Gear Manufacturers Association, Alexandria, VA (1973).

2. G. W. Michalec, *Precision Gearing Theory and Practice*, John Wiley and Sons, New York (1966).

3. *Gear and Spline Design* (DEP-1-6000 BKT), International Business Machines Corp., Armonk, NY (1970).

4. D. W. Dudley, Ed., *Gear Handbook*, McGraw-Hill, New York (1962).

11

Lubricants and Lubrication

Surface measurement of any metal gear tooth contact surface will indicate some degree of peaks and valleys. When gears are placed in mesh, irregular contact surfaces are brought together in the typical combination of rolling and sliding motion. The surface peaks, or asperities, of one tooth randomly contact the asperities of the mating tooth. Under the right conditions, the asperities form momentary welds that are broken off as the gear tooth action continues. Increased friction and higher temperatures, plus wear debris introduced into the system, are the result of this action.

The basic function of a lubricant is to provide an oil film that will separate two mating surfaces that move relative to one another. In metal gearing, it is imperative that an adequate lubricant and lubrication system be provided to prevent contact of surface asperities. Once failure of the lubricant or lubrication system is initiated, ultimate failure of the gearing is likely.

Plastics Reactions

In plastics gearing, both molded and cut plastics gears have the peak-and-valley surface contour. This is the result of manufacturing, inherent machining equipment inaccuracies, and allowable tolerances. Some studies indicate that under the right conditions, momentary welding can occur in plastics gears. A compressive stress is present as a set of gear teeth come into contact. The stress moves from the initial point of contact along the tooth profile until the teeth are no longer in contact. The compression causes the same subsurface stress as in metal gears. When relative sliding takes place at

297

the mating point of contact, heat builds up at a localized point and material is removed due to the shear stress. These factors contribute to

1. New, exposed surface irregularities
2. Free debris particles and erosion possibilities
3. Increased energy requirements to maintain constant speed
4. Increased friction and wear
5. Increased heat generation
6. Erratic and sluggish system response
7. Accelerated tooth contact surface change reflected in output load fluctuations or motion transfer problems

Significant lubrication differences and similarities are found between lubrication of metal and plastics gears. Applications, materials, and design situations range in plastic gearing from the extreme of plastics gearing with no lubrication and unfilled material to gears operating immersed in water, oil, or chemical baths. Present-day usage consists of many combinations of lube/no-lube, filled/unfilled materials, and like/unlike materials. The ideal low-cost gearing system is that requiring no lubrication and unfilled materials.

In a gear set that is designed, manufactured, assembled, and operated correctly, the use of a lubricant is recommended during the run-in period. Continued lubrication serves primarily to help reduce friction and assist in heat dissipation at the tooth contact surfaces, since even the best quality standard gears cannot avoid some degree of sliding contact during operation. Other uses of a lubricant in the application are for flushing wear particles, dirt, and moisture, providing corrosion protection to adjoining parts and lubrication of those parts. As in all plastics gearing applications, gears should be tested to determine design suitability. The lubricant and lubrication method should be tested at the same time using the identical systems of the intended application at the required service conditions.

Coefficient of friction, temperatures, stress level and wear factor of mating materials are an indication of the necessity for use of a lubricant. A low coefficient of friction indicates that relatively small amounts of input energy are necessary to overcome sliding contact conditions. Small wear factors for unit load will provide longer wear life. When coefficient of friction and wear data are not available, substitute materials may be considered. This is particularly advisable in plastics gearing because much data have been generated for the commonly used and most successful gearing materials. Temperature considerations have been covered in design sections of Chapter 8.

It is important to remember that lubricants are chemicals. Plastics are susceptible to chemical attack so a major consideration is the type of lubricant selected for a particular application. This selection process is aided by tables provided by plastics material suppliers and texts containing results of chemical compatibility tests. Test samples of candidate plastics materials are immersed in the chemical of interest at a certain temperature for a period of time. Test samples are then weighed and that weight compared with pre-test weights. Chemical attack of the plastic material has occurred if the sample weight has been reduced or if crazing of the material is evident. If the test sample weight is increased, the indication is that absorption has occurred. Remeasurement of the test sample can sometimes indicate the severity of the fluid absorption. In gearing, moisture or chemical absorption can be as severe a problem as chemical attack, because small clearances for backlash can easily be eliminated and wear initiated.

Discussion of chemical attack on plastics is not an indictment of the lubricant. Practically all types of lubricating oils contain at least one additive, and some oils contain several different types of additives. The amount of additive used varies from a few hundredths of a percent to 30 percent or more (1). It is usually the chemical action of the additives that is responsible for the failure of plastics materials when in contact for a period of time, under stress conditions, subjected to adverse temperatures, or in contact with combinations of other system materials.

Chemical compatibility data will usually indicate exposure time and temperatures. The question confronting the design engineer is the applicability of the data for his or her application. Operating stress levels are usually never the same as the stress level of the test sample. The same is true for the temperature and the time of exposure. For this reason, some material suppliers provide data generated at wide ranges of temperatures for extremely long periods of time. Regardless, gear life tests should always be run unless significant experience with a particular lubricant dictates otherwise.

Plastics Stress Level

The fact that lubricant attack is influenced by stress level is sometimes overlooked. Often samples submitted for chemical compatibility testing will be at a specific stress level due to normal sample preparation procedures. When a gear is produced either by molding or cutting, stresses are set up in the parts. These residual stresses may or may not be relieved with subsequent manufacturing processes. Nevertheless, operating stresses are also present during running of the gears in their application. The problem is that the reaction of the gear materials to the residual or operating stress levels and operating temperatures may produce significant lubricant and material incompatibilities.

Fortunately, experience has shown that some lubricants work better
with certain materials used in gears of common sizes, with typical
loads, and limited to reasonable temperature levels. The result is
that many material suppliers and gear houses are aware of the lu-
bricant/plastics gear compatibility concern and can be of assistance
in providing recommendations. However, since stress levels and applications can be substantially different, the recommendation is to
test the gears with the intended lubricant in actual situations. Where
testing is impossible or impractical, all available experience and re-
ported data should be consulted and analyzed.

Plastics gear lubrication is accomplished using the following methods or in combinations of the various methods.

1. Dry with no external or internal lubricant
2. Initial application of external lubricant, usually grease
3. Initial application, replenished at random or fixed intervals
4. Continuous coverage by liquid bath
5. Fillers such as carbon, graphite, or molybdenum-disulfide
6. Gears filled with lubricants such as silicone
7. Gears both filled with lubricants and externally lubricated

Other Considerations

A problem often encountered is adherence of the lubricant to the tooth-
contacting surfaces. Squeeze-out and throw-off by centrifugal action
has plagued gear users and is a continual problem in many applications. Some innovative housing designs have provided deflectors that
channel the oil or grease back into the gear contact area. Selection
of an adhering type lubricant may resolve the problem in some applications. Nonspreading and nonmigrating lubricants or oil creep bar-
rier films may also be possible if carefully selected for particular problems.

There are times when lubricants may be considered to be contami-
nants. This may be particularly true where the lubricant is used on
food handling equipment. Inadvertant contact with the food necessi-
tates the use of certain types such as the silicones.

Acetal (polyformaldehyde) is not vulnerable to solvation (attack by
lubricant components) or crazing. However, it is quite sensitive to
buildup of acidic constituents. The most popular gearing materials,
acetal and nylon, are susceptible to chemical attack at temperatures
above 150°F and in strong acids and strong alkalis, particularly at
full strength (3,4).

The most versatile synthetic lubricant families are the silicones
and hydrocarbons, where operating temperature ranges of -65°F to
+250°F are not uncommon.

Chen and Juarbe (5) discuss lubricants and MoS_2-filled nylon gears. Gear oils with an EP additive in the viscous range of 200-300cs at 40°C are suitable for nylon. This is equivalent to the AGMA mild EP lubricant #4 EP. Loads tested were heavy and low-speed operation.

Chemical equipment and chemical handling equipment can be sources of contamination by oils and greases. Lubricants can contaminate areas such as office equipment, where paper forms, bills, and account ledger materials must pass through data processing machines. Care is necessary so that creep, splash-out, dripping, or bleed do not become a problem.

What to Look for in a Plastics Lubricant

Items of importance are as follows:

1. *Correct viscosity*: Minimum oil film thickness, continual recreation of a lubricated surface, formation of a protective film, good distribution with minimum squeeze-out.
2. *Adequate temperature range*: Fluid film at low-temperature extreme, sufficient coverage and lubricating capability at high temperature extreme, minimum fluid breakdown at high temperature.
3. *Chemical stability*: Minimum oxidation under heat buildup may have additive protection.
4. *Good lubricity*: Minimum friction that aids in control of operating temperature rise may have additive protection.

Lubricant and Plastics Compatibility (2)

Materials not usually a problem are:

Nylon
Phenolic
Diallyl phthalate
Terephthalate polyesters
Polytetrafluoroethylene
Polyethylene
Polypropylene

Materials that can be a problem are:

Polystyrene
Polyvinyl chloride
ABS resins
Polycarbonate

Lubricants and Lubrication

Polysulfone
Polyphenylene oxides

References

1. J. G. Wills, *Lubrication Fundamentals*, Marcel Dekker, Inc., New York (1980).

2. *Nye Lubeletter*, William F. Nye, Inc., New Bedford, MA (1977).

3. *DuPont Delrin Acetal Resin Design Handbook*, E. I. DuPont DeNemours and Co., Wilmington, DE (1981).

4. *DuPont Zytel Nylon Resin Design Handbook*, Bulletin E-44971, E. I. DuPont DeNemours and Co., Wilmington, DE.

5. J. Chen, F. Juarbe, *Tests of MoS$_2$-Filled Nylon Gears Provide Design Data*, Power Transmission Design, Penton/IPC, Cleveland, OH (1978).

12

Drawing Specifications

The gear drawing is a clear representation of the part by sketch and dimensions as required in the end product configuration and quality level, but which does not describe manufacturing and measuring methods. Whenever possible, methods of making gears should be left to the discretion of the manufacturer, unless interchangeability or replacements necessitate the use of a certain method. Auxiliary views or sections may be given for clarification, but double-dimensioning and redundancy must be avoided.

Requirements

The drawing should contain part configuration, tooth data and possibly blank information, and notes that are specific and complete. Completeness is essential, but the drawing should only specify the minimum or basic number of items necessary to provide the correct part to its intended quality. Unnecessary data requirements increase costs because of the attention they must be given in tool design, mold design, manufacture of the tools, and inspection of the parts. The AGMA cautions that no detail essential to the operation of the gears should be omitted or assumed because the gear drawing may become, and usually is, a part of a contract between a gear manufacturer and a buyer (1).

To avoid both insufficient and unnecessary data, suggestions for notes, formats, and a checklist are provided for guidance. To insure that the gears will perform satisfactorily in unusual or critical conditions, the gear engineer should consult the checklist when selecting additional items of gear blank or tooth geometry.

With the versatility inherent in the molding process, it is common practice to combine two or more parts into one molded part. When orientation of surfaces or features is essential, views and notes should be added to depict the orientation clearly. For example, if a cam surface is part of a hub, it may be important to relate the rise or fall or provide angular relationship to a specific timing hole, mounting surface, or gear tooth. A cam chart may also be a requirement on the drawing. On the other hand, if specific orientation of cams, splines, keyways, or other special configurations are not a requirement, this should be stated in a note.

Certain gear tooth element requirements may be needed in special cases. These are usually handled by providing a special view of the tooth or feature and additional notes or by charts that describe those tooth elements, for example, profile or lead tolerances using charts as shown in Chapter 10. Gear blank features are sometimes shown to indicate the location surface in the application if not clearly indicated on the final part. This situation is a special case in molding plastics gears. Another view may be used to show a close-up of the gear teeth and indicate root, tooth form, start of true involute, and start of tip modification diameters. Tooth root, profile, and edge-round configurations, locations, and their tolerances can also be given in this view. Description of details of gear teeth by this method is fairly common.

Drawing conventions of individual companies will take precedence, but the AGMA recommends that notes should be grouped into one location on a drawing and the gear data in another location. Material designations are usually placed in the notes or in special designated blocks. Seldom, if ever, should the gear data and notes be intermixed.

A special case arises when a mounting shaft, bushing, or insert is pressed into a molded part or the bore machined after molding the gear teeth. In metal gearing, the teeth are cut with a topping hob or shaper so that the outside diameter of the gear teeth can be used as a reference surface and the bore or shaft center is held concentric to the gear teeth. In plastics gearing, the same concentricity must be maintained, primarily in the molding process. This type of gear assembly requires that permissible tolerance on concentricity of the gear teeth with surfaces machined or assembled after molding should be specified on the assembly drawing as "Maximum Composite Radial Variation" for inspection by running in tight mesh with a master gear on a variable-center-distance type of gage. Diametral pitch and pressure angle are also given to facilitate selection of the proper master gear. This applies to gears in the fine- to medium-pitch size range that will fit on the running gage. For coarse-pitch gears, concentricity must also be specified and inspected.

The above inspection method is a composite check of tooth thickness variation, spacing, and profile variations in addition to concentricity. Therefore the composite variation after assembly must be
equal to or slightly greater than that allowed on the component gear
before assembly. Spur or helical gear data shown on the assembled
gear drawing will be:

1. Number of teeth
2. Diametral pitch
3. Pressure angle
4. Maximum composite radial variation (when in tight mesh with master)

This design practice is considered a special case because of the un-
avoidable possibility that tolerance extremes can combine to cause
either slight interference or additional backlash.

Formats

Drawing formats, check lists, and notes for plastics gears are basically the same as those used for metal gears. Exceptions are material
specification, hardness, and finishes. Using plastics, the generic
material name is normally specified. Whenever requirements are such
that only a specific material has proven successful, the exact material
supplier designation is specified until other materials are qualified for
the application.

AGMA recommendations for gear specifications are provided as a
guide as they relate to metal gears. Only minimal modifications are
necessary to apply them to plastics gears. An additional requirement
that may sometimes be necessary is specification of a stress relaxation
treatment.

AGMA Standards Listing

The formats presented in Figures 12.1 through 12.18 are as recommended by the AGMA. They are arranged to show the data needed
on a part print, possible additional data for special applications, and
a listing showing the suggested number of decimal places. A listing
of applicable AGMA standards if further research is necessary in a
particular area of gear design and specification follows.

SPUR GEAR DATA				
Basic Specification Data	Number of teeth	(N)		
	Diametral pitch	(P)		
	Pressure angle	(φ)		
	Standard pitch diameter	(N/P)	(Ref.)	
	Tooth form			
	Max. calc. cir. thickness on std. pitch circle			
Mfg. and Inspection Data	Gear testing radius			
	AGMA quality number			
	Total composite tolerance			
	Tooth-to-tooth composite tolerance			
	Outside diameter			
	Master gear basic cir. tooth thickness at std. pitch circle		(Ref.)	
	Master gear number of teeth		(Ref.)	

HELICAL GEAR DATA[a]				
Basic Specification Data	Number of Teeth	(N)		
	Normal Diametral pitch	(P_n)		
	Normal Pressure angle	(ϕ_n)		
	Helix Angle – Hand	(ψ)		
	Standard pitch diameter	$(N/P_n \cos \psi)$	(Ref.)	
	Tooth form			
	Max. calc. normal cir. thickness on std. pitch circle			
Mfg. and Inspection Data	Gear testing radius			
	AGMA quality number			
	Total composite tolerance			
	Tooth-to-tooth composite tolerance			
	Lead			
	Outside diameter			
	Master gear basic normal cir. tooth thickness at std. pitch circle		(Ref.)	
	Master gear number of teeth		(Ref.)	

FIGURE 12.1   Recommended minimum spur and helical gear specifications for general applications.   If desired, a combination format covering both spur and helical gears can be used by specifying the helix angle equal to zero degress.   This permits standardization on the helical drawing format for both spur and helical gears.   (From Ref. 1.)

May be included in the drawing specification, listed in the process routine, or otherwise made available to manufacturing and inspection personnel, dependent on organization of the company involved.

GEAR PROCESSING DATA	
Measurement over two .XXXX diameter pins. For setup purposes only. Finished gear must satisfy drawing specification when checked with a master.	
Testing load (Ounces)	
Root diameter (Ref.)	

RECOMMENDED ADDITIONAL DATA FOR CONTROL GEARING

(Transmission of Angular Motion)

Add to mfg. and inspection data block

Mfg. and Inspection Data	Profile tolerance	
	Lead tolerance	
	Pitch Tolerance	

RECOMMENDED DATA FOR SPECIAL APPLICATIONS

In some cases, an end view of the tooth may be shown on which are indicated:

Surface texture of functional profile
Surface texture of fillet and root (if critical)
Functional profile diameter
Outside diameter
Pitch diameter
Root diameter
Form diameter
Base diameter
Tooth position angular tolerance
Runout tolerance

FIGURE 12.2 Recommended additional processing data for general purpose applications.

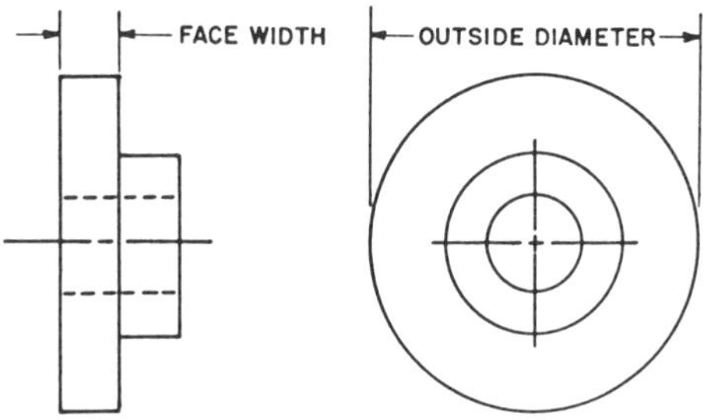

	Minimum Spur Gear Data	Minimum Helical Gear Data	Additional Optional Data	SPUR AND HELICAL GEAR DATA	Suggested Number of Decimal Places
Basic Specifications	X	X		Number of teeth	XXX
	X			Diametral pitch	XX
		X		Normal diametral pitch	XXX.XXXX
			X	Transverse diametral pitch	XXX.XXXX
	X			Pressure angle	XX.XXXX°
		X		Normal pressure angle	XX.XXXX°
			X	Transverse pressure angle	XX.XXXX°
		X		Helix angle	XX.XXXX°
		X		Hand of helix	LH or RH
	X	X		Standard pitch diameter (Ref.)	X.XXXX
	X	X		Tooth form	Full Depth (etc.)
			X	Addendum (Ref.)	.XXX
			X	Whole Depth (Ref.)	.XXX
	X			Max. calc. cir. thickness on std. pitch circle	.XXXX
		X		Max. calc. normal cir. thickness on std. pitch circle	.XXXX

FIGURE 12.3 Spur and helical gear specification information.

Minimum Spur Gear Data	Minimum Helical Gear Data	Additional Optional Data			
			SPUR AND HELICAL GEAR DATA		Suggested Number of Decimal Places
X	X		Gear testing radius		X.XXXX $^{+X.XXXX}_{-X.XXXX}$
X	X		AGMA quality number		QXX
X	X		Total composite tolerance		.XXXX
X	X		Tooth-to-tooth composite tolerance		.XXXX
X	X		Master specifications (numbers of teeth and tooth thickness or Drawing number)		XXX.XXX
		X	Testing load (ounces)		XX
		X	Measurement over two .XXXX diameter pins (for setup only)		X.XXXX $^{+.0000}_{-.00XX}$
	X		Lead		XXX.XXX
X	X		Outside diameter (preferably shown on drawing of gear)		X.XXX $^{+.000}_{-.00X}$
		X	Max. root diameter (external) min. root diameter (internal)		X.XXX
		X	Functional profile diameter		X.XXX
		X	Surface texture of functional profile		XX Micro in. (etc.)
		X	Profile tolerance		.00XX
		X	Lead tolerance		.00XX
		X	Circular pitch tolerance		.00XX
		X	Mating gear part number Dwg. no.		XXX.XXX
		X	Number of teeth in mating gear		XXX
		X	Minimum operating center distance		X.XXXX

Left margin labels: Manufacturing and Inspection; Engrg. Ref.

FIGURE 12.3 (continued)

BEVEL GEAR DATA (STRAIGHT TEETH)		
Number of teeth (N)		
Diametral pitch (P)		
Standard pitch diameter (N/P)	(Ref.)	
Pressure angle (∅)		
Shaft angle		
Pitch angle		
Addendum	(Ref.)	
Dedendum	(Ref.)	
Whole depth		
Root angle		
Face angle		
Tooth form		
Cir. thickness on pitch circle		
Number of teeth in mating gear		
Mating gear part number		
Quality control gear number		
Chordal addendum		
Chordal thickness		
AGMA quality number		
Backlash with mate at specified mounting distance		
Total composite tolerance		
Tooth-to-tooth composite tolerance		

(Left labels: Basic Specification; Mfg. and Inspection Data)

FIGURE 12.4 Recommended minimum straight tooth bevel gear specification.

May be included in the drawing specification, listed in the process routine or otherwise made available to manufacturing and inspection personnel, dependent on organization of the company involved.

I. GEAR PROCESSING DATA

		Member	Pinion	Gear
Gear Processing Data		Machine Setup Summary Number		
		Cutting or Grinding Distance		
	Cutter Specifications	Diameter		
		Point Width		
		Blade Radius		
		Pressure Angle (ϕ)		

RECOMMENDED ADDITIONAL DATA FOR CONTROL GEARING

(Transmission of Angular Motion)

Add to mfg. and inspection data block

Mfg. & Inspection Data		Initial Contact Pattern Area (Located at Center of Tooth)[a]		%
		Pitch Tolerance		

II. OPERATING DATA

Operating Data	Driving Member is:		
	Direction of Rotation (Looking at Back) CCW CW		
	Speed Range (RPM)		

III. MATERIAL AND HEAT TREATMENT DATA

Material and Heat Treatment Data	Material		
	Heat Treatment[b]		
	Depth of Case		
	Case Hardness		
	Core Hardness		

FIGURE 12.5 Recommended additional straight tooth bevel gear processing data.
[a]In reference to a mating member.
[b]A preliminary heat treatment may also be specified.

Minimum Straight Bevel Gear Data	Additional Optional Data		STRAIGHT BEVEL GEAR DATA		Suggested Number of Decimal Places
X			Number of teeth		XXX
X			Diametral pitch		XX
X			Pressure angle		XX.XXXX°
X			Shaft angle		XX.XXXX°
X			Standard pitch diameter	(Ref.)	X.XXXX
X			Pitch angle		XX.XXXX°
	X		Outer cone distance	(Ref.)	X.XXXX
X			Addendum	(Ref.)	.XXXX
X			Dedendum	(Ref.)	.XXXX
X			Whole depth		.XXX
X			Root angle		XX.XXXX°
X			Face angle		XX.XXXX°
X			Cir. thickness on standard pitch circle		.XXXX
X			Number of teeth in mating gear		XXX
X			Mating gear part number		(Dwg. No.)

(Left side label: Basic Specification and Engineering Data)

FIGURE 12.6   Straight bevel gear specification information.   (From Ref. 1.)

		STRAIGHT BEVEL GEAR DATA	Suggested Number of Decimal Places
	X	Testing load (ounces)	XX
X		AGMA quality number	QXX
X		Control gear number[a]	(Dwg. No.)
X		Chordal addendum[b]	.XXXX
	X	Chordal thickness[b]	.XXXX
	X	Tooth angle	XX.XXXX°
	X	Tool point width	.XXX
	X	Machine setup summary number	(Dwg. No.)
X		Backlash with mating gear at specified mounting distance	.00X to .00XX
X		Total composite tolerance	.00XX
X		Tooth-to-tooth composite tolerance	.00XX
	X	Initial Contact Pattern Area (located at center of tooth)	%

Column headers (left to right): Minimum Straight Bevel Gear Data; Additional Optional Data; (leftmost vertical label) Manufacturing and Inspection Data

FIGURE 12.6 (continued)
[a]Reference should be made to mating "quality control gear" or mating member.
[b]Not recommended for 48 pitch and finer.

	BEVEL GEAR DATA (SPIRAL[a], ZEROL[a] TEETH)		
BASIC SPECIFICATION	NUMBER OF TEETH (N)		
	DIAMETRAL PITCH (P)		
	PRESSURE ANGLE (ϕ)		
	SPIRAL ANGLE		
	STANDARD PITCH DIAMETER (N/P)	(Ref.)	
	SHAFT ANGLE		
	PITCH ANGLE		
	HAND OF SPIRAL – PINION		
	ADDENDUM	(Ref.)	
	DEDENDUM	(Ref.)	
	WHOLE DEPTH		
	ROOT ANGLE		
	FACE ANGLE		
	TOOTH FORM		
	CIR. THICKNESS ON STD. PITCH CIRCLE		
	NUMBER OF TEETH IN MATING GEAR		
	MATING GEAR PART NUMBER		
MANUFACTURING AND INSPECTION	QUALITY CONTROL GEAR PART NUMBER		
	AGMA QUALITY NUMBER		
	TOTAL COMPOSITE TOLERANCE		
	TOOTH-TO-TOOTH COMPOSITE TOLERANCE		
	MINIMUM TOOTH FILLET RADIUS		
	MACHINE SETUP SUMMARY NUMBER		
	BACKLASH WITH MATE AT SPECIFIED MOUNTING DISTANCE		
	TESTING LOAD (OUNCES)		
	INITIAL CONTACT PATTERN AREA (LOCATED AT CENTER OF TOOTH)		%

FIGURE 12.7 Recommended minimum spiral and Zerol-bevel gear specifications for general applications.
[a]One or other; not both should be shown.

I. GEAR PROCESSING DATA

GEAR PROCESSING DATA	MEMBER		PINION	GEAR
	MACHINE SETUP SUMMARY NUMBER			
	CUTTING OR GRINDING DISTANCE			
	CUTTER SPECIFICA-TIONS	DIAMETER		
		POINT WIDTH		
		BLADE RADIUS		
		PRESSURE ANGLE (ϕ)		

II. OPERATING DATA

OPERATING DATA	DRIVING MEMBER IS:		
	DIRECTION OF ROTATION (LOOKING AT BACK) CCW CW		
	SPEED RANGE (RPM)		

III. MATERIAL AND HEAT TREATMENT DATA

MATERIAL AND HEAT TREATMENT DATA	MATERIAL		
	HEAT TREATMENT[a]		
	DEPTH OF CASE		
	CASE HARDNESS		
	CORE HARDNESS		

FIGURE 12.8 Additional processing data for spiral and Zerol-bevel gears.
[a]A preliminary heat treatment may also be specified.

Minimum Bevel Gear Data	Additional Optional Data		SPIRAL AND ZEROL BEVEL GEAR DATA		Suggested Number of Decimal Places
X			Number of teeth		XXX
X			Diametral pitch		XX
	X		Standard pitch diameter	(Ref.)	X.XXXX
X			Pitch angle		XX.XXXX°
X			Shaft angle		XX.XXXX°
X			Pressure angle		XX.XXXX°
X			Spiral angle		XX.XXXX°
X			Hand of spiral – pinion		RH or LH
	X		Outer cone distance	(Ref.)	XX.XXXX
X			Addendum	(Ref.)	.XXXX
X			Dedendum	(Ref.)	.XXXX
X			Whole depth		.XXXX
X			Root angle		X.XXXX°
X			Face angle		X.XXXX°
X			Circular thickness – pinion		.XXXX
X			Circular thickness – gear		.XXXX

FIGURE 12.9 Spiral and Zerol-bevel gear specification information.
(From Ref. 1.)

Minimum Bevel Gear Data	Addational Optional Data	Explanation of Item Numbers		SPIRAL AND ZEROL BEVEL GEAR DATA		Suggested Number of Decimal Places
	X	29A	Normal chordal addendum	pinion	.XXXX
				gear[a]	.XXXX
	X	30A	Normal chordal thickness on standard pitch circle	pinion	.XXXX
				gear[a]	.XXXX
X		46	Mating gear part number		(Dwg. No.)
X		47	Number of teeth in mating gear		XXX
X		52	Backlash with mate at specified mounting distance		.00X to .00X
X		23a	Total composite tolerance		.00XX
X		24	Tooth-to-tooth composite tolerance		.00XX
X		22	AGMA quality number		QXX
X		50	Machine setup summary number[b]		(Summary Dwg. No.)
X		46A	Quality control gear number[c]		(Dwg. No.)
		X	26	Testing load (ounces)		XX

FIGURE 12.9 (continued)

[a]Omit for 48 DP and finer.

[b]Summary contains all tooth, machine setup, and cutter and grinding-wheel data.

[c]Reference should be made to mating "quality control gear" or mating member.

	WORM DATA		
BASIC SPECIFICATION	NUMBER OF THREADS IN WORM		
	AXIAL PITCH		
	LEAD ANGLE	(Ref.)	
	LEAD		
	NORMAL PRESSURE ANGLE		
	STANDARD PITCH DIAMETER	(Ref.)	
	HAND		
	TOOTH FORM		
	MAX. CALC. THREAD THICKNESS AT STD. PITCH CIRCLE (AXIAL)	(Ref.)	
MANUFACTURING AND INSPECTION DATA	MATING GEAR PART OR MASTER GEAR NUMBER		
	AGMA QUALITY NUMBER		
	GRINDING WHEEL OR CUTTER DIAMETER		
	TOOTH-TO-TOOTH COMPOSITE TOLERANCE		
	TOTAL COMPOSITE TOLERANCE		
	OUTSIDE DIAMETER		
	WORM TESTING RADIUS		

FIGURE 12.10 Recommended minimum worm specifications for general applications. In special cases, it may be desirable to inspect the worm for axial thread spacing (thread-to-thread and accumulative, over 3 axial pitches), or lead tolerance (over 1 axial pitch and over 3 axial pitches). These values are then shown in place of the master gear number, the tooth-to-tooth and total composite tolerance, and the worm testing radius. The recommended additional worm processing data shown in Figure 12.11 is then required.

> May be included in the drawing specification, listed in process routine, or otherwise made available to the manufacturing and inspection personnel, dependent on the organization of the company involved.

WORM PROCESSING DATA	
MEASUREMENT OVER 3.XXXX DIAMETER PINS[a] (WIRES)	
TESTING LOAD (OUNCES)	
MAX. ROOT DIAMETER	

FIGURE 12.11 Recommended additional worm processing data.
[a]If a master gear is used to check the worm, this item should include the note "For setup purposes only. Finished worm must conform to drawing specifications when checked with a master wormgear."

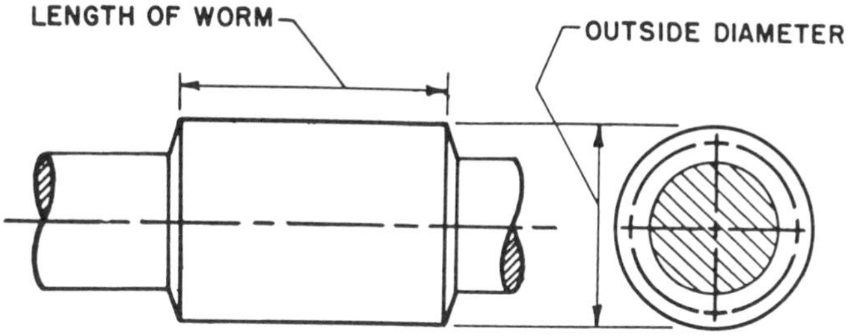

		WORM DATA		Suggested Number of Decimal Places
Minimum Specification for Worm / **Additional Optional Data** → columns				
Basic Specifications				
X		Number of threads in worm		XX
X		Axial pitch		.XXXX ±.XXXX
X		Lead angle	(Ref.)	XX.XXX
X		Lead		XX.XXXX ±.XXXX
X		Standard pitch diameter	(Ref.)	XX.XXXX
X		Hand		RH or LH
X		Normal Pressure Angle		XX.XXXX°
X		Tooth form		Std.
	X	Addendum	(Ref.)	.XXXX
	X	Whole depth	(Ref.)	.XXXX
X		Max. calc. thread thickness (axial)	(Ref.)	.XXXX
Manufacturing and Inspection Data				
X		Testing radius		XX.XXXX +.0000 −.XXXX
X		AGMA quality number		QXX
X		Total composite tolerance		.00XX
X		Tooth-to-tooth composite tolerance		.00XX
X		Master specifications[a]		(Dwg. No.)

FIGURE 12.12 Worm specification information. (From Ref. 1.)
[a]Reference should be made to mating "quality control gear" or mating member.

Minimum Specification for Worm	Additional Optional Data	WORM DATA	Suggested Number of Decimal Places
	X	Testing load (ounces)	XX
	X	Initial contact area	Central 50%
	X	Measurement over 3 .XXXX wires	X.XXXX to X.XXXX
	X	Normal chordal addendum[a]	.XXXX
	X	Normal chordal tooth thickness[a]	.XXXX
X		Outside diameter (preferably shown on drawing)	X.XXX $^{+.000}_{-.00X}$
	X	Maximum root diameter	X.XXX
	X	Functional profile diameter	X.XXX
	X	Surface texture of functional profile	XX micro inches
X		Lead tolerance	.XXXX
	X	Spacing tolerance	.XXXX
X		Grinding wheel or cutter diameter	XX
	X	Point width of grinding wheel or cutter (Ref.)	.XXX
	X	Pressure angle of wheel or cutter	XX.XXXX°
	X	Part number of mate	(Dwg. No.)
	X	Number of teeth in mating gear	XXX
X		Minimum backlash with mate at specified mounting distance (Ref.)	.XXXX

(Left margin labels: "Manufacturing and Inspection Data" spanning the upper rows; "Engineering Ref." spanning the last three rows.)

FIGURE 12.12 (continued)
[a]Not recommended for .065 axial pitch and finer.

WORMGEAR DATA		
BASIC SPECIFICA-TION	NUMBER OF TEETH	
	CIRCULAR PITCH (TRANSVERSE)	
	NORMAL PRESSURE ANGLE	
	LEAD ANGLE OF MATING WORM	
	TOOTH FORM	
MANUFACTURING AND INSPECTION DATA	GEAR TESTING RADIUS	
	AGMA QUALITY NUMBER	
	TOTAL COMPOSITE TOLERANCE	
	TOOTH-TO-TOOTH COMPOSITE TOLERANCE	
	OUTSIDE DIAMETER	
	MASTER WORM IDENTIFICATION	

RECOMMENDED ADDITIONAL PROCESSING DATA

May be included in the drawing specification, listed in the process routine, or otherwise made available to manufacturing and inspection personnel, dependent on organization of the company involved.

GEAR PROCESSING DATA	
TESTING LOAD (OUNCES)	
TOOL SPECIFICATIONS	

FIGURE 12.13 Recommended minimum wormgear specification for general applications.

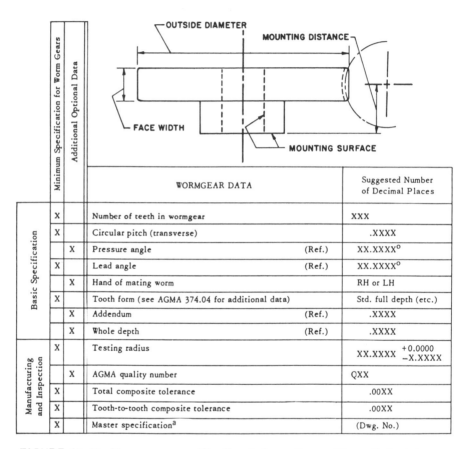

		WORMGEAR DATA		Suggested Number of Decimal Places
X		Number of teeth in wormgear		XXX
X		Circular pitch (transverse)		.XXXX
	X	Pressure angle	(Ref.)	XX.XXXX°
X		Lead angle	(Ref.)	XX.XXXX°
	X	Hand of mating worm		RH or LH
X		Tooth form (see AGMA 374.04 for additional data)		Std. full depth (etc.)
	X	Addendum	(Ref.)	.XXXX
	X	Whole depth	(Ref.)	.XXXX
X		Testing radius		XX.XXXX $^{+0.0000}_{-X.XXXX}$
	X	AGMA quality number		QXX
X		Total composite tolerance		.00XX
X		Tooth-to-tooth composite tolerance		.00XX
X		Master specification[a]		(Dwg. No.)

FIGURE 12.14 Wormgear specification information. (From Ref. 1.)
[a]Reference should be made to a mating control gear or mating gear part number.

		WORM GEAR DATA	Suggested Number of Decimal Places
	X	Testing load (ounces)	XX
	X	Initial contact area	Central 50%
X		Outside diameter (preferably shown on drawing)	X.XXX $^{+.000}_{-.00X}$
	X	Surface texture of functional profile	XX micro inches
	X	Spacing tolerance	.XXXX max.
	X	Mounting distance	X.XXXX
	X	Part number of mate	(Dwg. No.)
	X	Number of threads in mating worm	XX
	X	Minimum backlash with mate at specified mounting distance (Ref.)	.XXX

Left-side column labels: Minimum Specification for Worm Gears / Additional Optional Data; row groups: Manufacturing and Inspection, Engineering Ref.

FIGURE 12.14 (continued)

FACE GEAR DATA		
BASIC SPECIFICATION — NUMBER OF TEETH		
DIAMETRAL PITCH		
PRESSURE ANGLE (CUTTER)		
MOUNTING DISTANCE	(Ref.)	
CUTTER NUMBER		
WHOLE DEPTH	(Ref.)	
MANUFACTURING AND INSPECTION DATA — TESTING DIMENSION (CENTER LINE OF PINION TO MS)		
AGMA QUALITY NUMBER		
TOTAL COMPOSITE TOLERANCE		
TOOTH-TO-TOOTH COMPOSITE TOLERANCE		
OUTER DIAMETER		
INNER DIAMETER		
MASTER IDENTIFICATION		
MASTER NUMBER OF TEETH		

RECOMMENDED MINIMUM FINE-PITCH PINIONS SPECIFICATIONS FOR USE WITH FACE GEARS

For pinions to be used with face gears use the format for spur gears (for general applications)

RECOMMENDED ADDITIONAL PROCESSING DATA FOR FACE GEARS

May be included in the drawing specification, listed in the process routine, or otherwise made available to manufacturing and inspection personnel, dependent on organization of the company involved.

GEAR PROCESSING DATA	
TESTING LOAD (OUNCES)	
TOOTH CONTACT PATTERN DISPLACEMENT	

FIGURE 12.15 Recommended minimum face gear specification.

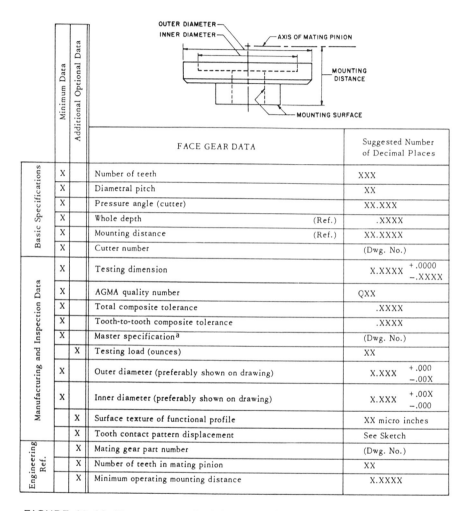

			FACE GEAR DATA		Suggested Number of Decimal Places
Basic Specifications	X		Number of teeth		XXX
	X		Diametral pitch		XX
	X		Pressure angle (cutter)		XX.XXX
	X		Whole depth	(Ref.)	.XXXX
	X		Mounting distance	(Ref.)	XX.XXXX
	X		Cutter number		(Dwg. No.)
Manufacturing and Inspection Data	X		Testing dimension		X.XXXX $^{+.0000}_{-.XXXX}$
	X		AGMA quality number		QXX
	X		Total composite tolerance		.XXXX
	X		Tooth-to-tooth composite tolerance		.XXXX
	X		Master specification[a]		(Dwg. No.)
		X	Testing load (ounces)		XX
	X		Outer diameter (preferably shown on drawing)		X.XXX $^{+.000}_{-.00X}$
	X		Inner diameter (preferably shown on drawing)		X.XXX $^{+.00X}_{-.000}$
		X	Surface texture of functional profile		XX micro inches
		X	Tooth contact pattern displacement		See Sketch
Engineering Ref.		X	Mating gear part number		(Dwg. No.)
		X	Number of teeth in mating pinion		XX
		X	Minimum operating mounting distance		X.XXXX

FIGURE 12.16   Face gear and pinion specification information.   (From Ref. 1.)

[a]Reference should be made to mating "quality control gear," master gear, or mating member.   (From Ref. 1.)

		SPUR RACK DATA	
BASIC SPECIFICATION DATA		NUMBER OF TEETH (IN TOTAL RACK LENGTH)	
		TRANSVERSE PITCH	
		PRESSURE ANGLE	
		TOOTH FORM	
		MAX. CALC. THICKNESS ON STD. PITCH PLANE	
		ADDENDUM (Ref.)	
		WHOLE DEPTH (Ref.)	
MFG. AND INSPECTION DATA		PITCH	
		AGMA QUALITY NUMBER	
		INDEX	
		PROFILE	
		LEAD	

		HELICAL RACK DATA	
BASIC SPECIFICATION DATA		NUMBER OF TEETH	
		NORMAL PITCH	
		NORMAL PRESSURE ANGLE	
		HELIX ANGLE	
		HAND OF HELIX	
		TOOTH FORM	
		MAX. CALC. NORMAL THICKNESS ON STD. PITCH PLANE	
		ADDENDUM (Ref.)	
		WHOLE DEPTH (Ref.)	
MFG. AND INSPECTION DATA		PITCH	
		AGMA QUALITY NUMBER	
		INDEX	
		PROFILE	
		LEAD	

FIGURE 12.17 Recommended minimum spur and helical rack specifications for general applications.

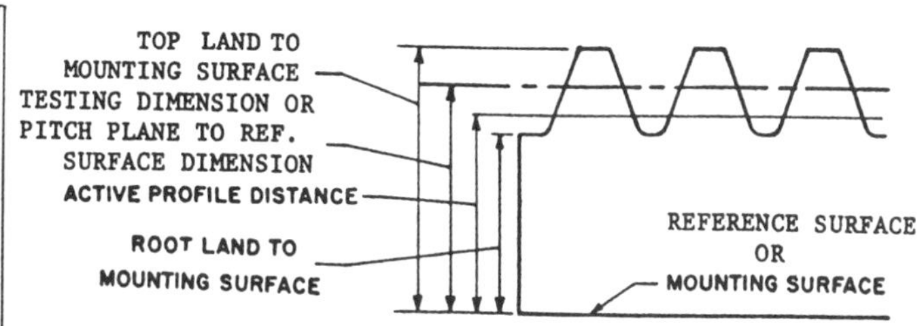

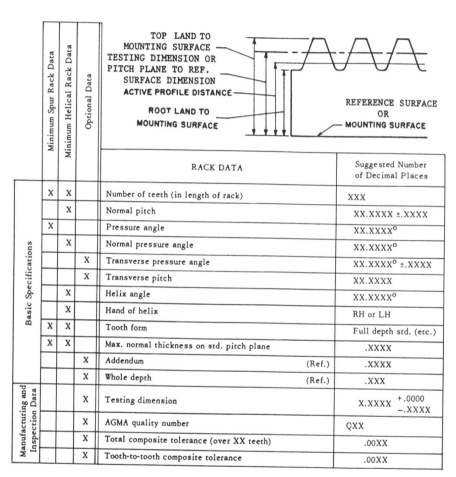

	Minimum Spur Rack Data	Minimum Helical Rack Data	Optional Data	RACK DATA		Suggested Number of Decimal Places
Basic Specifications	X	X		Number of teeth (in length of rack)		XXX
		X		Normal pitch		XX.XXXX ±.XXXX
	X			Pressure angle		XX.XXXX°
		X		Normal pressure angle		XX.XXXX°
			X	Transverse pressure angle		XX.XXXX° ±.XXXX
			X	Transverse pitch		XX.XXXX
		X		Helix angle		XX.XXXX°
		X		Hand of helix		RH or LH
	X	X		Tooth form		Full depth std. (etc.)
	X	X		Max. normal thickness on std. pitch plane		.XXXX
			X	Addendum	(Ref.)	.XXXX
			X	Whole depth	(Ref.)	.XXX
Manufacturing and Inspection Data			X	Testing dimension		X.XXXX $^{+.0000}_{-.XXXX}$
			X	AGMA quality number		QXX
			X	Total composite tolerance (over XX teeth)		.00XX
			X	Tooth-to-tooth composite tolerance		.00XX

FIGURE 12.18 Spur and helical rack specification information.

	Minimum Spur Gear Data	Minimum Helical Gear Data	Optional Data	RACK DATA	Suggested Number of Decimal Places
Manufacturing and Inspection Data	X	X		Master specification	(Dwg. No.)
			X	Testing load (ounces)	XX
			X	Measurement over 1 .XXXX pin	X.XXXX to X.XXXX
	X	X		Top land to mounting surface (shown on view)	X.XXXX $^{+.000}_{-.00X}$
			X	Root land to mounting surfaces	X.XXX $^{+.000}_{-.00X}$
			X	Functional profile distance	X.XXXX max.
			X	Surface texture of functional profile	XX micro inches
			X	Mating gear part number	(Dwg. No.)
			X	Number of teeth in mating gear	XXX
			X	Minimum operating distance to center line of mate	X.XXXX
	X	X		Pitch plane to reference surface dimension[a]	±.XXXX XXX.XXXX
Optional Data [a]				Testing dimension	
				Total composite tolerance (over XX teeth)	
				Tooth-to-tooth tolerance	

FIGURE 12.18 (continued)

[a]If composite action tolerances are specified, they should be in lieu of pitch, index, pitch plane to reference surface, lead, and profile inspections. Composite action inspection usually requires special fixtures. If pitch, index, pitch plane to reference surface, lead, and profile tolerances are specified, they should be in lieu of composite action tolerances.

AGMA Publications*

The following is a listing of AGMA standards and publications that relate to the format items by gear type.

Gear type	Area of concern	Standard number
Spur and helical	Tooth form	207.05, USAS B6.7-1967
	Tooth proportions	207.05
	Inspection	390.03
	Quality number	390.03
	Runout tolerance	390.03
Straight bevel	Tooth form	208.02
	Tooth proportions	208.02, 330.01
	Dimensioning methods	330.01
	Inspection	331.01, 390.03
	Quality number	331.01, 390.03
Spiral and Zerol	Tooth form	202.03, 209.03
	Tooth proportions	202.03, 209.03
	Inspection	331.01, 390.03
Worm	Tooth form	374.04
	Tooth proportions	374.04
Wormgear	Testing radius	370.01
	Tooth form	374.04
	Inspection	390.03
Face	Tooth proportions	203.03
Spur and helical rack	Tooth form	207.05

*See the precautionary note in the Introduction when referring to
AGMA documents.

References

1. *AGMA Gear Handbook, Volume 1: Gear Classification, Materials and Measuring Methods for Unassembled Gears*, American Gear Manufacturers Association, Alexandria, VA (1973).

Appendix A

Cosine, Tangent, and Involute of Some Selected Angles

Angle (degree)	Cosine	Tangent	Involute
14.0	0.9702957	0.2493280	0.0049819
14.1	0.9698720	0.2511826	0.0050912
14.2	0.9694453	0.2530389	0.0052021
14.3	0.9690157	0.2548968	0.0053147
14.4	0.9685832	0.2567564	0.0054289
14.5	0.9681476	0.2586176	0.0055448
14.6	0.9677092	0.2604805	0.0056624
14.7	0.9672678	0.2623451	0.0057817
14.8	0.9668234	0.2642114	0.0059027
14.9	0.9663761	0.2660794	0.0060254
15.0	0.9659258	0.2679492	0.0061498
15.1	0.9654726	0.2698207	0.0062760
15.2	0.9650165	0.2716940	0.0064039
15.3	0.9645574	0.2735690	0.0065337
15.4	0.9640954	0.2754459	0.0066652
15.5	0.9636305	0.2773245	0.0067985
15.6	0.9631626	0.2792050	0.0069337
15.7	0.9626917	0.2810873	0.0070706
15.8	0.9622180	0.2829715	0.0072095
15.9	0.9617413	0.2848575	0.0073501
16.0	0.9612617	0.2867454	0.0074927
16.1	0.9607792	0.2886352	0.0076372
16.2	0.9602937	0.2905269	0.0077835

(continued)

Angle (degree)	Cosine	Tangent	Involute
16.3	0.9598053	0.2924205	0.0079318
16.4	0.9593140	0.2943160	0.0080820
16.5	0.9588197	0.2962135	0.0082342
16.6	0.9583226	0.2981129	0.0083883
16.7	0.9578225	0.3000144	0.0085444
16.8	0.9573195	0.3019178	0.0087025
16.9	0.9568136	0.3038232	0.0088626
17.0	0.9563048	0.3057307	0.0090247
17.1	0.9557930	0.3076402	0.0091889
17.2	0.9552784	0.3095517	0.0093551
17.3	0.9547608	0.3114653	0.0095234
17.4	0.9542403	0.3133810	0.0096937
17.5	0.9537170	0.3152988	0.0098662
17.6	0.9531907	0.3172187	0.0100407
17.7	0.9526615	0.3191407	0.0102174
17.8	0.9521294	0.3210649	0.0103963
17.9	0.9515944	0.3229912	0.0105773
18.0	0.9510565	0.3249197	0.0107604
18.1	0.9505157	0.3268504	0.0109458
18.2	0.9499721	0.3287833	0.0111333
18.3	0.9494255	0.3307184	0.0113231
18.4	0.9488760	0.3326557	0.0115151
18.5	0.9483237	0.3345953	0.0117094
18.6	0.9477684	0.3365372	0.0119059
18.7	0.9472103	0.3384813	0.0121048
18.8	0.9466493	0.3404278	0.0123059
18.9	0.9460854	0.3423765	0.0125093
19.0	0.9455186	0.3443276	0.0127151
19.1	0.9449489	0.3462810	0.0129232
19.2	0.9443764	0.3482368	0.0131336
19.3	0.9438010	0.3501950	0.0133465
19.4	0.9432227	0.3521556	0.0135617
19.5	0.9426415	0.3541186	0.0137794
19.6	0.9420575	0.3560840	0.0139994
19.7	0.9414705	0.3580518	0.0142220
19.8	0.9408808	0.3600222	0.0144470
19.9	0.9402881	0.3619949	0.0146744
20.0	0.9396926	0.3639702	0.0149044
20.1	0.9390943	0.3659480	0.0151369
20.2	0.9384930	0.3679284	0.0153719
20.3	0.9378889	0.3699112	0.0156094

(continued)

Angle (degree)	Cosine	Tangent	Involute
20.4	0.9372820	0.3718967	0.0158495
20.5	0.9366722	0.3738847	0.0160922
20.6	0.9360595	0.3758753	0.0163375
20.7	0.9354440	0.3778685	0.0165854
20.8	0.9348257	0.3798644	0.0168359
20.9	0.9342045	0.3818629	0.0170891
21.0	0.9335804	0.3838640	0.0173449
21.1	0.9329535	0.3858679	0.0176034
21.2	0.9323238	0.3878744	0.0178646
21.3	0.9316912	0.3898837	0.0181286
21.4	0.9310558	0.3918957	0.0183953
21.5	0.9304176	0.3939105	0.0186647
21.6	0.9297765	0.3959280	0.0189369
21.7	0.9291326	0.3979483	0.0192119
21.8	0.9284858	0.3999715	0.0194897
21.9	0.9278363	0.4019974	0.0197703
22.0	0.9271839	0.4040262	0.0200538
22.1	0.9265286	0.4060579	0.0203401
22.2	0.9258706	0.4080924	0.0206293
22.3	0.9252097	0.4101299	0.0209215
22.4	0.9245460	0.4121703	0.0212165
22.5	0.9238795	0.4142136	0.0215145
22.6	0.9232102	0.4162598	0.0218154
22.7	0.9225381	0.4183091	0.0221193
22.8	0.9218632	0.4203613	0.0224262
22.9	0.9211854	0.4224165	0.0227361
23.0	0.9205049	0.4244748	0.0230491
23.1	0.9198215	0.4265361	0.0233651
23.2	0.9191353	0.4286005	0.0236842
23.3	0.9184464	0.4306680	0.0240063
23.4	0.9177546	0.4327386	0.0243316
23.5	0.9170601	0.4348124	0.0246600
23.6	0.9163627	0.4368893	0.0249916
23.7	0.9156626	0.4389693	0.0253264
23.8	0.9149597	0.4410526	0.0256642
23.9	0.9142540	0.4431390	0.0260053
24.0	0.9135455	0.4452287	0.0263497
24.1	0.9128342	0.4473216	0.0266973
24.2	0.9121201	0.4494178	0.0270481
24.3	0.9114033	0.4515173	0.0274023
24.4	0.9106837	0.4536201	0.0277598

(continued)

Angle (degree)	Cosine	Tangent	Involute
24.5	0.9099613	0.4557263	0.0281206
24.6	0.9092361	0.4578357	0.0284848
24.7	0.9085082	0.4599486	0.0288523
24.8	0.9077775	0.4620649	0.0292232
24.9	0.9070440	0.4641845	0.0295976
25.0	0.9063078	0.4663077	0.0299753
25.1	0.9055688	0.4684342	0.0303566
25.2	0.9048271	0.4705643	0.0307413
25.3	0.9040825	0.4726978	0.0311295
25.4	0.9033353	0.4748349	0.0315213
25.5	0.9025853	0.4769755	0.0319166
25.6	0.9018325	0.4791197	0.0323154
25.7	0.9010770	0.4812675	0.0327179
25.8	0.9003188	0.4834189	0.0331239
25.9	0.8995578	0.4855739	0.0335336
26.0	0.8987940	0.4877326	0.0339470
26.1	0.8980276	0.4898949	0.0343640
26.2	0.8972584	0.4920610	0.0347847
26.3	0.8964864	0.4942308	0.0352092
26.4	0.8957118	0.4964043	0.0356374
26.5	0.8949344	0.4985816	0.0360694
26.6	0.8941542	0.5007627	0.0365051
26.7	0.8933714	0.5029476	0.0369447
26.8	0.8925858	0.5051363	0.0373881
26.9	0.8917975	0.5073290	0.0378354
27.0	0.8910065	0.5095254	0.0382866
27.1	0.8902128	0.5117259	0.0387416
27.2	0.8894164	0.5139302	0.0392006
27.3	0.8886172	0.5161385	0.0396636
27.4	0.8878154	0.5183508	0.0401306
27.5	0.8870108	0.5205671	0.0406015
27.6	0.8862036	0.5227874	0.0410765
27.7	0.8853936	0.5250117	0.0415555
27.8	0.8845810	0.5272402	0.0420387
27.9	0.8837656	0.5294727	0.0425259
28.0	0.8829476	0.5317094	0.0430172
28.1	0.8821269	0.5339503	0.0435128
28.2	0.8813035	0.5361953	0.0440124
28.3	0.8804774	0.5384445	0.0445163
28.4	0.8796486	0.5406980	0.0450245
28.5	0.8788171	0.5429557	0.0455369

(continued)

Angle (degree)	Cosine	Tangent	Involute
28.6	0.8779830	0.5452177	0.0460535
28.7	0.8771462	0.5474840	0.0465745
28.8	0.8763067	0.5497547	0.0470998
28.9	0.8754645	0.5520297	0.0476295
29.0	0.8746197	0.5543091	0.0481636
29.1	0.8737722	0.5565929	0.0487020
29.2	0.8729221	0.5588811	0.0492450
29.3	0.8720693	0.5611738	0.0497924
29.4	0.8712138	0.5634710	0.0503442
29.5	0.8703557	0.5657728	0.0509006
29.6	0.8694949	0.5680791	0.0514616
29.7	0.8686315	0.5703899	0.0520271
29.8	0.8677655	0.5727054	0.0525973
29.9	0.8668967	0.5750255	0.0531721
30.0	0.8660254	0.5773503	0.0537515
30.1	0.8651514	0.5796797	0.0543356
30.2	0.8642748	0.5820139	0.0549245
30.3	0.8633956	0.5843528	0.0555181
30.4	0.8625137	0.5866965	0.0561164
30.5	0.8616292	0.5890450	0.0567196
30.6	0.8607420	0.5913984	0.0573276
30.7	0.8598523	0.5937565	0.0579405
30.8	0.8589599	0.5961196	0.0585582
30.9	0.8580649	0.5984877	0.0591809
31.0	0.8571673	0.6008606	0.0598086
31.1	0.8562671	0.6032386	0.0604412
31.2	0.8553643	0.6056215	0.0610788
31.3	0.8544588	0.6080095	0.0617215
31.4	0.8535508	0.6104026	0.0623692
31.5	0.8526402	0.6128008	0.0630221
31.6	0.8517269	0.6152041	0.0636801
31.7	0.8508111	0.6176126	0.0643432
31.8	0.8498927	0.6200263	0.0650116
31.9	0.8489717	0.6224452	0.0656851
32.0	0.8480481	0.6248694	0.0663640
32.1	0.8471219	0.6272988	0.0670481
32.2	0.8461932	0.6297336	0.0677376
32.3	0.8452618	0.6321738	0.0684324
32.4	0.8443279	0.6346193	0.0691326
32.5	0.8433914	0.6370703	0.0698383
32.6	0.8424524	0.6395267	0.0705493

(continued)

Angle (degree)	Cosine	Tangent	Involute
32.7	0.8415108	0.6419886	0.0712659
32.8	0.8405666	0.6444560	0.0719880
32.9	0.8396199	0.6469290	0.0727157
33.0	0.8386706	0.6494076	0.0734489
33.1	0.8377187	0.6518918	0.0741878
33.2	0.8367643	0.6543817	0.0749324
33.3	0.8358074	0.6568772	0.0756826
33.4	0.8348479	0.6593785	0.0764385
33.5	0.8338858	0.6618856	0.0772003
33.6	0.8329212	0.6643984	0.0779678
33.7	0.8319541	0.6669171	0.0787411
33.8	0.8309845	0.6694417	0.0795204
33.9	0.8300123	0.6719721	0.0803055
34.0	0.8290376	0.6745085	0.0810966
34.1	0.8280603	0.6770509	0.0818936
34.2	0.8270806	0.6795993	0.0826967
34.3	0.8260983	0.6821537	0.0835058
34.4	0.8251135	0.6847143	0.0843210
34.5	0.8241262	0.6872810	0.0851424
34.6	0.8231364	0.6898538	0.0859699
34.7	0.8221440	0.6924328	0.0868036
34.8	0.8211492	0.6950181	0.0876435
34.9	0.8201519	0.6976097	0.0884898
35.0	0.8191520	0.7002075	0.0893423
35.1	0.8181497	0.7028118	0.0902012
35.2	0.8171449	0.7054224	0.0910665
35.3	0.8161376	0.7080395	0.0919382
35.4	0.8151278	0.7106630	0.0928165
35.5	0.8141155	0.7132931	0.0937012
35.6	0.8131008	0.7159297	0.0945925
35.7	0.8120835	0.7185729	0.0954904
35.8	0.8110638	0.7212227	0.0963949
35.9	0.8100416	0.7238793	0.0973061
36.0	0.8090170	0.7265425	0.0982240
36.1	0.8079899	0.7292125	0.0991487
36.2	0.8069603	0.7318894	0.1000802
36.3	0.8059283	0.7345730	0.1010185
36.4	0.8048938	0.7372636	0.1019637
36.5	0.8038569	0.7399611	0.1029159
36.6	0.8028175	0.7426655	0.1038750
36.7	0.8017756	0.7453770	0.1048412

(continued)

Angle (degree)	Cosine	Tangent	Involute
36.8	0.8007314	0.7480956	0.1058144
36.9	0.7996847	0.7508212	0.1067947
37.0	0.7986355	0.7535541	0.1077822
37.1	0.7975839	0.7562941	0.1087769
37.2	0.7965299	0.7590413	0.1097788
37.3	0.7954735	0.7617959	0.1107880
37.4	0.7944146	0.7645577	0.1118046
37.5	0.7933533	0.7673270	0.1128285
37.6	0.7922896	0.7701037	0.1138599
37.7	0.7912235	0.7728878	0.1148987
37.8	0.7901550	0.7756795	0.1159451
37.9	0.7890841	0.7784788	0.1169990
38.0	0.7880108	0.7812856	0.1180605

Appendix B
Glossary of Gearing Terms

The following were taken from *Gear Nomenclature — Terms, Definitions, Symbols, and Abbreviations*, American Gear Manufacturers Association (AGMA 112.04), Alexandria, VA (1976).

1.0 General Designations

1.01 *Gears* are machine elements that transmit motion by means of successively engaging teeth (Figure B.1).

1.02 A *gear* is a machine part with gear teeth. Of two gears that run together, the one with the larger number of teeth is called the gear (Figure B.1).

1.03 A *pinion* is a gear with a small number of teeth. Of two gears that run together, the one with the smaller number of teeth is called the pinion (Figure B.1).

1.04 A *rack* is a gear with teeth spaced along a straight line and suitable for straight-line motion. A *basic rack* is one that is adopted as the basis of a system of interchangeable gears. Standard gear-tooth proportions are often illustrated on an outline of the basic rack (Figure B.1). A *generating rack* is a rack outline used to indicate tooth details and dimensions for the design of a required generating tool, such as a hob or a gear-shaper cutter.

1.05 A *worm* is a gear with one or more teeth in the form of screw threads (Figures B.2 and B.8).

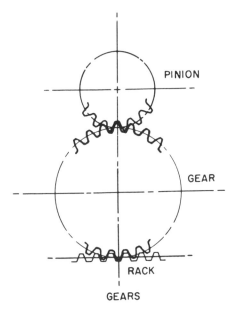

PINION

GEAR

RACK

GEARS

FIGURE B.1

2.0 Kinds of Gears

2.01   *Spur gears* are cylindrical in form and operate on parallel axis. The teeth are straight and parallel to the axes (Figure B.3).

2.02   A *spur rack* has straight teeth that are at right angles to the direction of motion (Figure B.3).

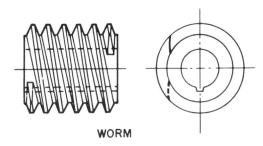

WORM

FIGURE B.2

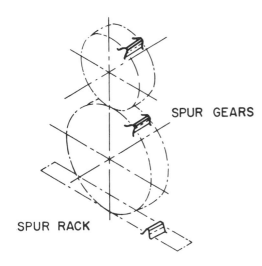

SPUR GEARS

SPUR RACK

FIGURE B.3

2.03 A *helical gear* is cylindrical in form and has helical teeth (Figure B.4).

2.04 *Parallel helical gears* operate on parallel axes and, when both are external, the helices are of opposite hand (Figure B.5).

2.05 *Crossed helical gears* operate on crossed axes and may have teeth of the same or of opposite hand. The term *crossed helical gears* has superseded the old term *spiral gears* (Figure B.6).

LEFT-HAND
HELICAL TOOTH

HELICAL GEAR

FIGURE B.4

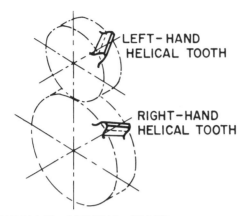

PARALLEL HELICAL GEARS

FIGURE B.5

2.06   *Single-helical gears* have teeth of only one hand on each gear
        (Figure B.7).
2.07   *Double-helical gears* each have both right-hand and left-hand
        helical teeth and operate on parallel axes. These are also known
        as herringbone gears (Figure B.7).
2.08   A *helical rack* has straight teeth that are oblique to the direc-
        tion of motion (Figure B.7).

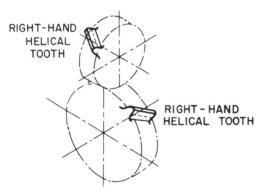

CROSSED HELICAL GEARS

FIGURE B.6

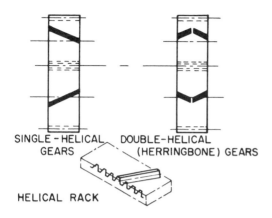

SINGLE-HELICAL GEARS   DOUBLE-HELICAL (HERRINGBONE) GEARS

HELICAL RACK

FIGURE B.7

2.09　A *cylindrical worm* has one or more threads in the form of screw threads on a cylinder (Figure B.2).

2.10　An *hourglass worm* has one or more threads and increases in diameter from its middle portion toward both ends, conforming to the curvature of the gear. It is sometimes called an enveloping worm (Figure B.9).

2.11　A *wormgear* is the mate to a worm. A wormgear that is completely conjugate to its worm has line contact and is said to be single enveloping (Figure B.8). It is usually cut by a tool that

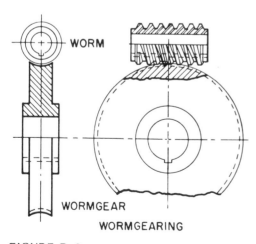

WORM

WORMGEAR

WORMGEARING

FIGURE B.8

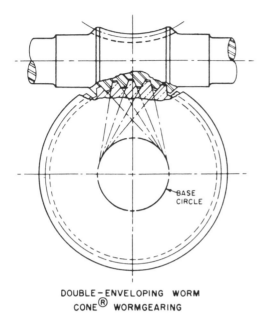

DOUBLE-ENVELOPING WORM
CONE® WORMGEARING

FIGURE B.9

is geometrically similar to the worm. An involute spur gear
or helical gear used with a cylindrical worm has only point
contact.

2.12   *Wormgearing* includes worms and their mating gears. The axes
are usually at right angles (Figure B.8).

2.13   *Double-enveloping wormgearing* comprises hourglass worms
mated with fully conjugate worm-gears (Figure B.9).

2.14   *Bevel gears* are conical in form and operate on intersecting
axes, which are usually at right angles (Figure B.10).

2.15   *Miter gears* are mating bevel gears with equal numbers of teeth
and with axes at right angles (Figure B.11).

2.16   *Angular bevel gears* are bevel gears in which the axes are not
at right angles (Figure B.11).

2.17   A *crown gear* is a bevel gear having a plane pitch surface. It
corresponds in bevel gears to the rack in spur gears (Figure
B.12).

2.18   *Straight bevel gears* have straight tooth elements, which if extended, would pass through the point of intersection of their
axes (Figure B.13).

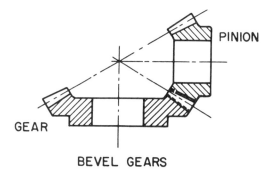

BEVEL GEARS

FIGURE B.10

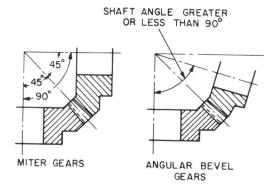

FIGURE B.11

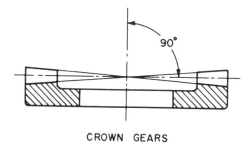

FIGURE B.12

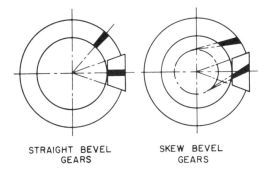

STRAIGHT BEVEL   SKEW BEVEL
     GEARS   GEARS

FIGURE B.13

2.19   *Spiral bevel gears* have teeth that are curved and oblique (Figure B.14).

2.20   *Zerol bevel gears* have teeth that are curved but in the same general direction as straight teeth.   They are spiral bevel gears of zero spiral angle (Figure B.14).

2.21   *Skew bevel gears* are those for which the corresponding crown gear has teeth that are straight and oblique (Figure B.13).

2.22   *Hypoid gears* are similar in general form to bevel gears but operate on nonintersecting axes (Figure B.15).

2.23   *Face gears* consist of a spur or helical pinion in combination with a conjugate gear of disk form, the axes usually being at right angles, either intersecting or nonintersecting (Figure B.16).

2.24   An *external gear* is one with the teeth formed on the outer surface of a cylinder or cone (Figure B.17).

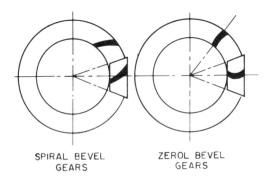

SPIRAL BEVEL   ZEROL BEVEL
     GEARS   GEARS

FIGURE B.14

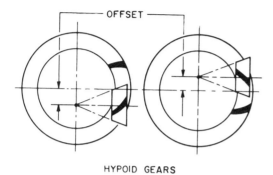

HYPOID GEARS

FIGURE B.15

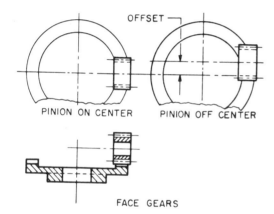

PINION ON CENTER   PINION OFF CENTER

FACE GEARS

FIGURE B.16

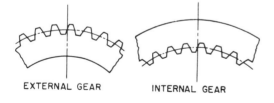

EXTERNAL GEAR   INTERNAL GEAR

INTERNAL BEVEL GEAR

FIGURE B.17

Appendix B

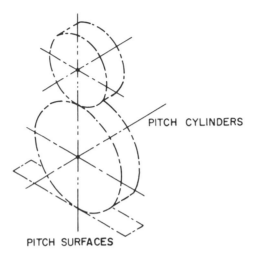

PITCH CYLINDERS

PITCH SURFACES

FIGURE B.18

2.25   An *internal gear* is one with the teeth formed on the inner sur-
       face of a cylinder cone (Figure B.17).   An internal gear can be
       meshed only with an external pinion.
2.26   *Special gears* include Beveloid, Coniflex®, Formate®, Helicon,
       Helixform®, Revacycle®, Spheroid, Spiroid, and Zerol.

3.0 Pitch Surfaces

The following are for gears having a constant ratio of angular ve-
locities and either parallel or intersecting axes and therefore not

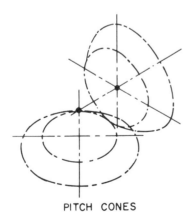

PITCH CONES

FIGURE B.19

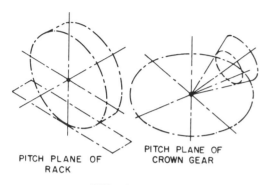

PITCH PLANE OF
RACK

PITCH PLANE OF
CROWN GEAR

PITCH PLANES

FIGURE B.20

including crossed helical gears, wormgearing, hypoid gears, or offset face gears.

3.01   *Pitch surfaces* are the imaginary planes, cylinders, or cones that roll together without slipping. For a constant-velocity ratio, the pitch cylinders and pitch cones are circular (Figures B.18 and B.19).

3.02   A *pitch plane* is the imaginary surface in a rack or in a crown gear that rolls without slipping with a pitch cylinder or pitch cone of another gear (Figures B.20 and B.21).

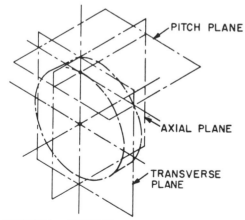

PITCH PLANE

AXIAL PLANE

TRANSVERSE
PLANE

PRINCIPAL   REFERENCE   PLANES

FIGURE B.21

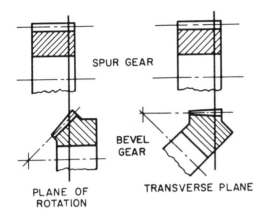

PLANE OF
ROTATION

TRANSVERSE PLANE

FIGURE B.22

3.03   A *pitch cylinder* is the imaginary cylinder in a gear that rolls
without slipping on a pitch cylinder or pitch plane of another
gear (Figure B.18).

3.04   A *pitch cone* is the imaginary cone in a bevel gear that rolls
without slipping on a pitch cone or pitch plane of another gear
(Figure B.19).

4.0 Principal Planes

4.01   The *axial plane* of a pair of gears is the plane that contains
the two axes. In a single gear, an axial plane may be any
plane containing the axis and a given point (Figure B.21).

4.02   The *pitch plane* of a pair of gears is the plane perpendicular
to the axial plane and tangent to the pitch surfaces. A pitch
plane in an individual gear may be any plane tangent to its
pitch surface. The pitch plane of a rack or crown gear is the
pitch surface (Figure B.21).

4.03   A *plane of rotation* is any plane perpendicular to a gear axis
(Figure B.22).

4.04   A *transverse plane* is perpendicular to the axial plane and to
the pitch plane. In gears with parallel axes, the transverse
plane and plane of rotation coincide (Figures B.21 and B.22).

4.05   A *tangent plane* is tangent to the tooth surfaces at a point or
line of contact (Figures B.23 and B.24).

4.06   A *normal plane* is in general normal to a tooth surface at a
pitch point and perpendicular to the pitch plane. In a helical rack, a normal plane is normal to all the teeth it intersects.

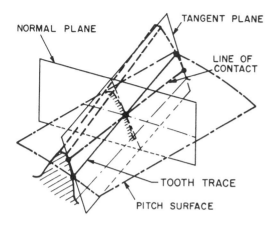

NORMAL PLANE

TANGENT PLANE

LINE OF CONTACT

TOOTH TRACE

PITCH SURFACE

FIGURE B.23

In a helical gear, however, a plane can be normal to only one tooth at a point lying in the plane surface. At such a point, the normal plane contains the line normal to the tooth surface and is normal to the pitch surface (Figure B.23).

4.07 The *principal reference planes* are a pitch plane, an axial plane, and a transverse plane, all intersecting at a point and mutually perpendicular (Figure B.21).

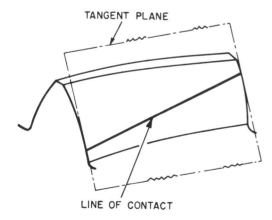

TANGENT PLANE

LINE OF CONTACT

FIGURE B.24

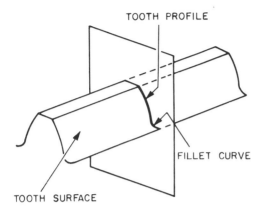

TOOTH PROFILE

FILLET CURVE

TOOTH SURFACE

FIGURE B.25

5.0 Elements of Gear Teeth

5.01   The *tooth surface* forms the side of a gear tooth (Figure B.25).

5.02   A *tooth profile* is one side of a tooth in a cross section between the outside circle and the root circle. Usually a profile is the curve of intersection of a tooth surface and a plane or surface normal to the pitch surface, such as the transverse, normal, or axial plane (Figure B.25).

5.03   The *fillet curve* is the concave portion of the tooth profile where it joins the bottom of the tooth space (Figure B.25).

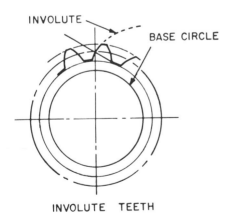

INVOLUTE

BASE CIRCLE

INVOLUTE   TEETH

FIGURE B.26

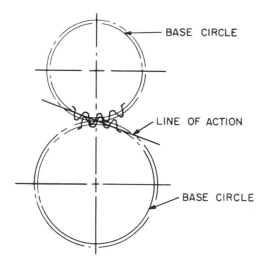

FIGURE B.27

5.04 *Involute teeth* of spur gears, helical gears, and worms, are those in which the profile in a transverse plane (exclusive of the fillet curve) is the involute of a circle (Figure B.26).

5.05 The *base circle* is the circle from which involute tooth profiles are derived (Figures B.26 and B.27).

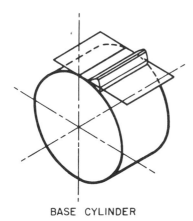

BASE CYLINDER

FIGURE B.28

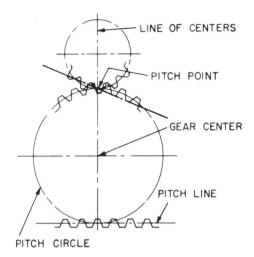

LINE OF CENTERS

PITCH POINT

GEAR CENTER

PITCH LINE

PITCH CIRCLE

FIGURE B.29

5.06   The *base cylinder* corresponds to the base circle and is the cylinder from which involute tooth surfaces, either straight or helical, are derived (Figure B.28).

5.07   A *pitch circle* is the curve of intersection of a pitch surface of revolution and a plane of rotation. According to theory, it is the imaginary circle that rolls without slipping with a pitch circle of a mating gear (Figure B.29).

5.08   The *pitch line* corresponds in the cross section of a rack to the pitch circle in the cross section of a gear (Figure B.29).

5.09   A *gear center* is the center of the pitch circle (Figure B.29).

5.10   The *line of centers* connects the centers of the pitch circles of two engaging gears; it is also the common perpendicular of the axes in crossed helical gears and wormgears. When one of the gears is a rack, the line of centers is perpendicular to its pitch line (Figure B.29).

5.11   The *pitch point* is the point of tangency of two pitch circles (or of a pitch circle and pitch line) and is on the line of centers. The pitch point of a tooth profile is at its intersection with the pitch circle (Figure B.29).

5.12   The *addendum circle* coincides with the tops of the teeth in a cross section (Figure B.30).

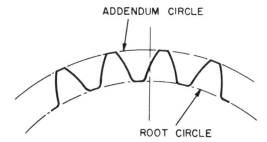

FIGURE B.30

5.13 The *root circle* is tangent to the bottom of the tooth spaces in a cross section (Figure B.30).

5.14 A *point of contact* is any point at which two tooth profiles touch each other (Figure B.31).

5.15 The *path of action* is the imaginary curve along which contact occurs during the engagement of two tooth profiles. It is the trace of the surface of action in the surface containing the profiles (Figures B.31 and B.32).

5.16 The *line of action* is the path of action for involute gears. It is the straight line passing through the pitch point and tangent to the base circle (Figure B.27).

5.17 A *line of contact* is the line or curve along which two tooth surfaces are tangent to each other (Figures B.23, B.24, and B.32).

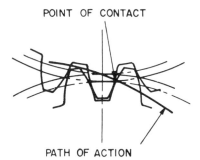

FIGURE B.31

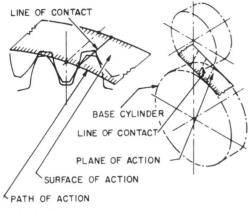

LINE OF CONTACT

BASE CYLINDER
LINE OF CONTACT
PLANE OF ACTION
SURFACE OF ACTION
PATH OF ACTION

FIGURE B.32

6.0 Linear and Circular Dimensions

6.01   *Center distance* (C) is the distance between the parallel axes
        of spur gears or of parallel helical gears, or the crossed axes
        of crossed helical gears or of worms and wormgears. Also, it
        is the distance between the centers of the pitch circles (Fig-
        ure B.33).

6.02   *Pitch* (p) is the distance between similar, equally spaced tooth
        surfaces along a given line or curve. The use of the single

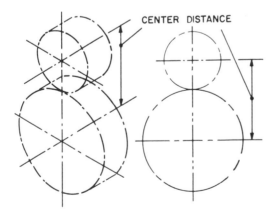

CENTER DISTANCE

FIGURE B.33

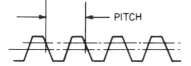

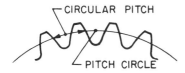

FIGURE B.34

word pitch without qualification may be confusing, and for
this reason, specific designations are preferred, like circular
pitch, axial pitch, diametral pitch (Figure B.34).

6.03 *Circular pitch* (p) is the distance along the pitch circle or
pitch line between corresponding profiles of adjacent teeth
(Figure B.34).

6.04 *Base pitch* (P_b) in an involute gear is the pitch on the base
circle or along the line of action. Corresponding sides of involute gear teeth are parallel curves, and the base pitch is
the constant and fundamental distance between them along a
common normal in a plane of rotation (Figure B.35).

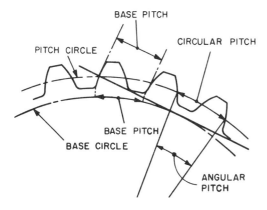

FIGURE B.35

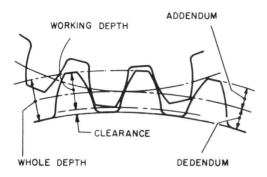

FIGURE B.36

6.05   *Addendum* (a) is the height by which a tooth projects beyond
        the pitch circle or pitch line and also the radial distance be-
        tween the pitch circle and the addendum circle (Figure B.36).
6.06   *Dedendum* (b) is the depth of a tooth space below the pitch
        circle or pitch line and also the radial distance between the
        pitch circle and the root circle (Figure B.36).
6.07   *Clearance* (c) is the amount by which the dedendum in a given
        gear exceeds the addendum of its mating gear (Figure B.36).
6.08   *Working depth* (h_k) is the depth of engagement of two gears,
        that is, the sum of their operating addendums (Figure B.36).
6.09   *Whole depth* (h_t) is the total depth of a tooth space equal to
        addendum plus dedendum and also equal to working depth
        plus clearance (Figure B.36).
6.10   *Pitch diameter* (D, d) is the diameter of the pitch circle (Fig-
        ure B.37).

        A *standard pitch diameter* is one calculated according to the
        standard pitch of the gear-cutting tool (see also 7.01).

        Operating pitch diameters are the pitch diameters determined
        from the numbers of teeth and the center distance at which
        gears operate. *Testing pitch diameter* is a value that may be
        used in the testing or inspection of a gear, and which should
        be defined in the testing procedure. In a bevel gear, the
        pitch diameter is understood to be at the outer ends of the
        teeth unless otherwise specified.

6.11   *Outside diameter* (D_o, d_o) is the diameter of the addendum
        (outside) circle. In a bevel gear it is the diameter of the

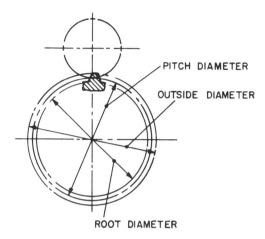

PITCH DIAMETER

OUTSIDE DIAMETER

ROOT DIAMETER

FIGURE B.37

crown circle. In a throated wormgear it is the maximum diameter of the blank (Figures B.37 and B.38). The term applies
to external gears.

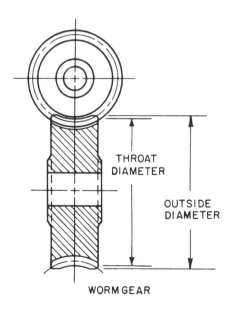

THROAT
DIAMETER

OUTSIDE
DIAMETER

WORMGEAR

FIGURE B.38

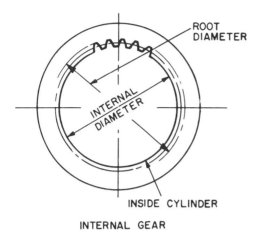

INTERNAL GEAR

FIGURE B.39

6.12   *Root diameter* (D_R, d_R) is the diameter of the root circle (Figures B.37 and B.39).

6.13   *Internal diameter* (d_i) is the diameter of the addendum circle of an internal gear (Figure B.39).

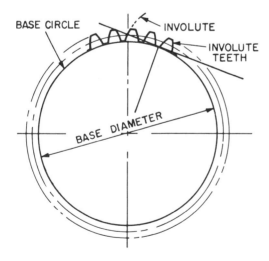

FIGURE B.40

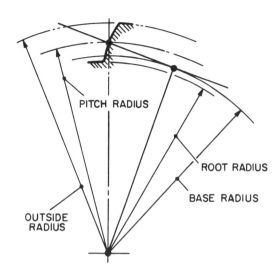

FIGURE B.41

6.14   *Base diameter* (D_b, d_b) is the diameter of the base cylinder of an involute gear (Figure B.40).

6.15   *Pitch radius* (R, r) is the radius of the pitch circle (Figure B.41).

6.16   *Outside radius* (R_o, r_o) is the radius of the addendum circle of an external gear (Figure B.41).

6.17   *Root radius* (R_R, r_R) is the radius of the root circle (Figure B.41).

6.18   *Base radius* (R_b, r_b) is the radius of the base circle of involute profiles (Figure B.41).

6.19   *Circular thickness* (t_G, t_P) is the length of arc between the two sides of a gear tooth on the pitch circle unless otherwise specified (Figure B.42).

6.20   *Base circular thickness* (t_b) in involute teeth is the length of arc on the base circle between the two involute curves forming the profiles of a tooth (Figure B.42).

6.21   *Chordal thickness* (t_c) is the length of the chord subtending a circular-thickness arc. Any convenient measuring diameter may be selected, not necessarily the pitch diameter (Figure B.42).

6.22   *Chordal addendum* (a_c) is the height from the top of the tooth to the chord subtending the circular-thickness arc. Any convenient measuring diameter may be selected, not necessarily the pitch diameter (Figure B.42).

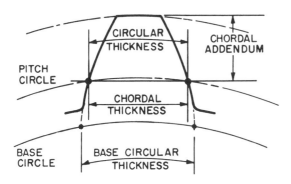

FIGURE B.42

6.23   *Backlash* (B) is the amount by which the width of a tooth space exceeds the thickness of the engaging tooth on the operating pitch circles (Figure B.43).

6.24   *Face width* (F) is the length of the teeth in an axial plane (Figure B.44).

6.25   *Effective face width* (Fe) is the portion that may actually come into contact with mating teeth, as occasionally one member of a pair of gears may have a greater face width than the other (Figure B.44).

6.26   *Total face width* (Ft) is the actual dimension of a gear blank that exceeds the effective face width or as in double-helical gears where the total face width includes any distance separating right-hand and left-hand helices (Figure B.44).

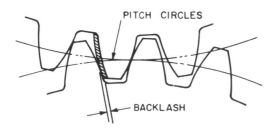

FIGURE B.43

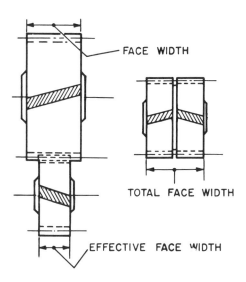

FIGURE B.44

6.27   *Length of action* (Z) is the distance on an involute line of action through which the point of contact moves during the action of the tooth profile (Figure B.45).

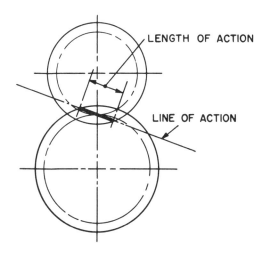

FIGURE B.45

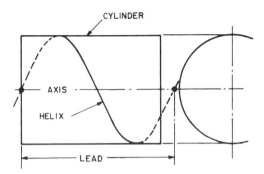

FIGURE B.46

6.28   *Lead* (L) is the axial advance of a helix for one complete turn,
        as in the threads of cylindrical worms and the teeth of helical
        gears (Figure B.46).

7.0 Angular Dimensions

7.01   *Pressure angle* (ϕ) is, in general, the angle at a pitch point
        between the line of pressure that is normal to the tooth sur-
        face, and the plane tangent to the pitch surface. The pres-
        sure angle gives the direction of the normal to a tooth profile
        (Figure B.47).

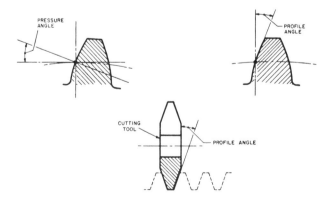

FIGURE B.47

Profile angle (ϕ) is, in general, the angle at a specified pitch point between a line tangent to a tooth surface and the line normal to the pitch surface (which is a radial line of a pitch circle). This definition is applicable to every type of gear for which a pitch surface can be defined. The symbol ϕ is used for both pressure angle and profile angle, because the two angles have the same magnitude. The profile angle gives the direction of the tangent to a tooth profile (Figure B.47).

In spur gears and straight bevel gears, tooth profiles are considered only in a transverse plane, and the general terms profile angle and pressure angle are customarily used rather than transverse profile angle and transverse pressure angle. In helical teeth, the profiles may be considered in different planes, and in specifications it is essential to use terms that indicate the direction of the plane in which the profile angle or the pressure angle lies, such as transverse profile angle, normal pressure angle, or axial profile angle.

Standard profile angles are established in connection with standard proportions of gear teeth and standard gear-cutting tools. Involute gears operate together correctly after a change of center distance, and gears designed for a changed center distance can be generated correctly by standard tools. A change in center distance is accompanied by changes in operating values for pitch diameter, circular pitch, diametral pitch, pressure angle, and tooth thickness or backlash. The same involute gear may be used under conditions that change its operating pitch diameter and pressure angle. Unless there is a good reason for doing otherwise, it is practical to consider that the pitch diameter and the profile angle of a single gear correspond to the pitch and the profile angle of the hob or cutter used to generate its teeth.

The operating pressure angle is determined by the base circles of two gears and the center distance at which the gears operate. Various other pressure angles may be considered in gear calculations.

In tools and gages for cutting, grinding, and gaging gear teeth, the profile angle is the angle between a cutting edge or a cutting surface, and some principal direction such as that of a shank, an axis, or a plane of rotation (Figure B.47).

7.02 *Helix angle* (ψ) is the angle between any helix and an element of its cylinder. In helical gears and worms, it is at the pitch diameter unless otherwise specified (Figure B.48).

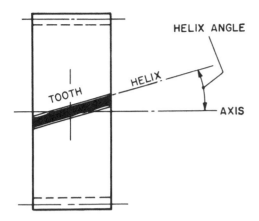

FIGURE B.48

7.03   *Lead angle* (λ) is the angle between any helix and a plane of
        rotation. It is the complement of the helix angle and is used
        for convenience in worms and hobs. It is understood to be
        at the pitch diameter unless otherwise specified (Figure B.49).
        In screw-thread practice the term helix angle was formerly
        used instead of the term lead angle.

7.04   *Shaft angle* (Σ) is the angle between the axes of two nonpar-
        allel gear shafts. In a pair of crossed helical gears, the
        shaft angle lies between the oppositely rotating portions of
        two shafts. This applies also in the case of wormgearing. In
        bevel gears, the shaft angle is the sum of the two pitch
        angles. In hypoid gears, the shaft angle is given when

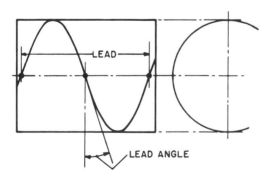

FIGURE B.49

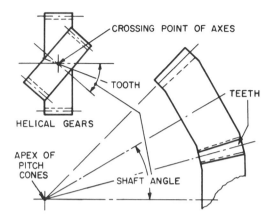

FIGURE B.50

starting a design, and it does not have a fixed relation to the pitch angles and spiral angles (Figure B.50).

7.05 *Involute polar angle* (θ) is the angle between a radius vector to a point on an involute curve and a radial line to the intersection of the curve with the base circle (Figure B.51).

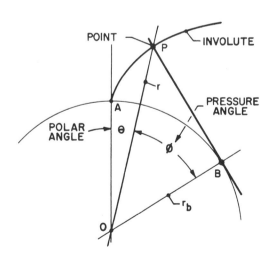

FIGURE B.51

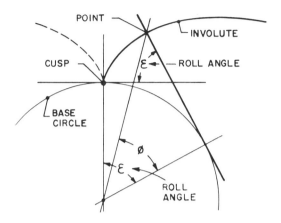

FIGURE B.52

7.06   *Involute roll angle* (ε) is the angle whose arc on the base circle
        of radius unity equals the tangent of the pressure angle at a
        selected point on the involute (Figure B.52).

8.0 Numbers and Ratios

8.01   *Number of teeth or threads* (N, n) is the number of teeth con-
        tained in the whole circumference of the pitch circle.

8.02   *Gear ratio* (m_G) is the ratio of the larger to the smaller number
        of teeth in a pair of gears.

8.03   *Diametral pitch* (P, P_d) is the ratio of the number of teeth to
        the pitch diameter in inches. Diametral pitch equals π divided
        by circular pitch (P = π/p). (For standard diametral pitch,
        see 6.10 and 7.01.)

8.04   *Angular pitch* (θ_N) is the angle subtended by the circular
        pitch, usually expressed in radians. N = (360/N) degrees,
        or (2π/N) radians (Figure B.35).

8.05   *Module (inch)* (m) is the ratio of the pitch diameter in inches
        to the number of teeth. It is the reciprocal of the diametral
        pitch.

8.06   *Module (millimeter)* (m) is the ratio of the pitch diameter in
        millimeters to the number of teeth. It is converted to (inch)
        diametral pitch by the equation:

$$Pd = \frac{25.4}{\text{module (millimeters)}}$$

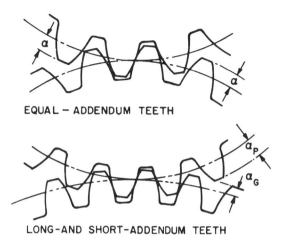

EQUAL – ADDENDUM TEETH

LONG-AND SHORT-ADDENDUM TEETH

FIGURE B.53

9.0 Miscellaneous Terms and Dimensions

9.01 *Full-depth teeth* are those in which the working depth equals 2.000 divided by normal diametral pitch.

9.02 *Stub teeth* are those in which the working depth is less than 2.000 divided by normal diametral pitch.

9.03 *Equal-addendum teeth* are those in which two engaging gears have equal addendums (Figure B.53).

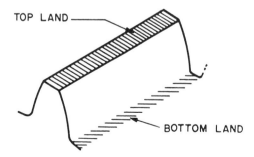

TOP LAND

BOTTOM LAND

FIGURE B.54

HOB OR TOOL

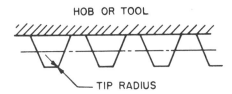

TIP RADIUS

FIGURE B.55

9.04   *Long- and short-addendum teeth* are those in which the adden-
        dums of two engaging gears are unequal (Figure B.53).
9.05   *Bottom land* is the surface at the bottom of a tooth space ad-
        joining the fillet (Figure B.54).
9.06   *Top land* is the surface of the top of a tooth (Figure B.54).
9.07   *Tip radius* (r_T) is the radius of the circular arc used to join
        a side-cutting edge and an end-cutting edge in gear-cutting
        tools. *Edge radius* is an alternate term (Figure B.55).
9.08   *Fillet radius* (r_f) is the radius of a circular arc approximating
        the fillet curve. In generated teeth, the fillet curve has a
        varying radius of curvature (Figure B.56).
9.09   *Undercut* is a condition in generated gear teeth when any part
        of the fillet curve lies inside of a line drawn tangent to the
        working profile at its point of juncture with the fillet. Under-
        cut may be deliberately introduced to facilitate finishing oper-
        ations (Figure B.57).
9.10   *Arc of action* (Q_t) is the arc of the pitch circle through which
        a tooth profile moves from the beginning to the end of contact
        with a mating profile (Figure B.58).
9.11   *Arc of approach* (Q_a) is the arc of the pitch circle through
        which a tooth profile moves from its beginning of contact un-
        til the point of contact arrives at the pitch point (Figure B.58).

FILLET RADIUS

PROFILE RADIUS OF
CURVATURE

FIGURE B.56

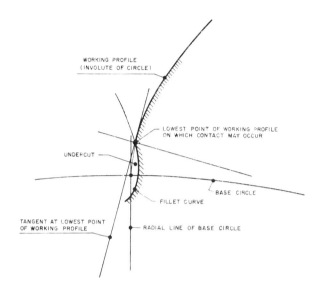

WITH UNDERCUT: THE FILLET CURVE INTERSECTS THE WORKING PROFILE
WITH NO UNDERCUT: THE FILLET CURVE AND THE WORKING PROFILE HAVE A COMMON TANGENT.

FIGURE B.57

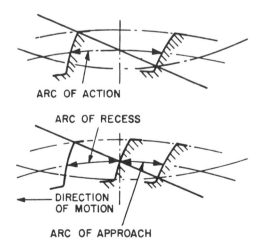

FIGURE B.58

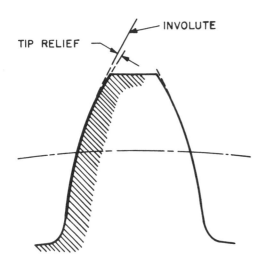

TIP RELIEF

INVOLUTE

FIGURE B.59

9.12   *Arc of recess* (Q_r) is the arc of the pitch circle through which
        a tooth profile moves from contact at the pitch point until con-
        tact ends (Figure B.58).

9.13   *Contact ratio* (m_g) in general is the number of angular pitches
        through which a tooth surface rotates from the beginning to
        the end of contact.

9.14   *Tip relief* is an arbitrary modification of a tooth profile where-
        by a small amount of material is removed near the tip of the
        gear tooth (Figure B.59).

The following were taken from *AGMA Gear Handbook, Volume 1: Gear
Classification, Materials and Measuring Methods for Unassembled Gears*
(AGMA 390.03) American Gear Manufacturers Association, Alexandria,
VA (1973).

1. *Axial runout* is runout measured in a direction parallel to the axis
    of rotation.

2. *Composite action* is the variation in center distance when two gears
    are rolled in tight mesh.

3. *Pitch range* is the difference between the longest and shortest
    pitches on a gear.

4. *Tooth-to-tooth composite action* is the greatest amount of compos-
    ite action within any single pitch or contact throughout an entire
    gear.

5. *Total composite action* is the total amount of composite action for
    an entire gear.

Appendix C
Glossary of Plastics Terms

The terms in Appendix C are used in molded and machined gearing and blanks (1).

1.1 Molding Terms

Automatic molding: A method of injection, compression, or transfer molding that repeatedly goes through the entire cycle without human assistance.

Cavity: Usually the female portion of the mold that forms the outer surface of the molded part. Depending on the number of such cavities, molds are designated as single cavity, two cavity, four cavity, etc., molds. The number of cavities corresponds to the number of parts made simultaneously in one machine.

Combination mold: See Family mold.

Compression molding: A technique of plastics molding in which the molding compound (generally preheated) is placed in the open mold cavity, the mold is closed under pressure (usually in a hydraulic press), causing the material to flow and fill the cavity completely, the pressure being held until the material has cured or solidified.

Core (V): To remove mass from a molded part, or to channel a mold for circulation of water, oil, or steam to maintain proper mold temperature.

Core (N): A pin or other part of a mold that forms a recess or hole in a molded part or a hole or channel in the mold used as a heat transfer medium.

Coring (molded part design): The removal of excess material from the cross section of a molded part to attain a more uniform wall thickness and reduce mass.

Cycle: The complete repeating sequence of operations in a process or part of a process. In molding, the cycle time is the period, or elapsed time, between a certain point in one cycle and the same point in the next cycle.

Ejector pin or ejector sleeve: A rod, pin, or sleeve that pushes a molded part off the force or core or out of the cavity of the mold. It is usually attached to an ejector bar or plate that can be actuated by the ejector mechanism of the press, or by auxiliary means, such as hydraulic or air cylinders, or springs.

Electrical discharge machining (EDM): A metal-working process, applicable to mold construction, in which controlled sparking from an electrical generator is used to remove unwanted material. The shape to be removed may be controlled by either an electrode or a wire that is programmed to follow the desired outline.

Electrode: A master used in a conventional electrical discharge machine. In making gear cavities, an electrode is cut in the shape of the gear to be molded, with all the necessary shrinkage factors taken into consideration. This electrode is then sunk into a metal blank to form a gear cavity.

Electroformed mold: A mold made by electroplating metal onto a master to form a thick shell that, when properly backed up and fitted, becomes a cavity or force. Usually hard nickel alloy is used.

Family mold: A multi-cavity mold wherein each of the cavities forms one of the component parts of the finished object; often applied to molds wherein parts for different customers are grouped together in one mold for economy of production.

Force: The portion of the mold forming the inside of the molded part (sometimes called a core).

Gate: A restricted orifice between the runner system and the gear cavity.

Injection molding: The process of forming a shape by forcing a material under pressure, from a heated cylinder through a sprue and runner system into the cavity of a closed mold.

Insert: Usually a metallic shape inserted into the cavity, around which the plastic material is molded, so that the insert becomes an integral part of the final product. Threaded inserts place accurate metallic threads in accurate locations within the molded part. A metal sleeve may be used to form the bore of a molded gear. A full metallic insert may be used so that only the rim and teeth of a gear are molded.

Mold (N): A hollow form into which a plastics material is placed or forced, and which imparts the final shape to the finished article.

Mold (V): To shape plastics parts in a mold by heat and pressure.

Molding cycle: The period of time used to complete the sequence of operations on a molding press, requisite for the production of one set of moldings (see Cycle).

Molding shrinkage (mold shrinkage, shrinkage, contraction): The difference in dimensions, usually expressed in inches per inch, between a molding and the cavity in which it was molded, both the mold and the molding being at normal room temperature when measured.

Parting line: Separation line between the two halves of the mold.

Runner system: The term usually applied to all the material in the form of sprues, runners, and gates that lead material from the nozzle of an injection machine or the pot of a transfer mold to the mold cavities.

Sprue: Feed opening, usually tapered for spur removal, provided in the injection or transfer mold as a passage from the machine nozzle to the runner system; also, the slug formed at this hole. Spur is a shop term for the sprue slug.

Transfer molding: A method of molding thermosetting materials, in which the plastic is first softened by heat and pressure in a transfer chamber, then forced by high pressure through suitable sprues, runners, and gates into a closed mold for final curing.

1.2 Processing Terms

Aging: The change of a material with time under defined environmental conditions, leading to improvement or deterioration of properties.

Anneal (for plastics): To heat a molded plastics article to a pre-determined temperature below its melting point, and slowly cool it, to relieve stresses. (Molded or machined plastics parts may be annealed dry in an oven, or wet in a heated tank of mineral oil or other heat transfer medium.)

Boss: Projection on a plastic part, designed to add strength, to facilitate alignment during assembly, to provide for fastenings, etc.

Cast: To form a "plastic" object by pouring a fluid monomer or polymer into an open mold, where it polymerizes or solidifies.

Cure: To change the physical properties of a material by chemical reaction. It may be by condensation, polymerization, or by vulcanization; usually accomplished by the action of heat and catalysts, with pressure.

Deflashing: Removal of the flash (excess, unwanted material) on a plastic molding by filing, sanding, milling, tumbling, abrading, wheelabrating, etc.

Ultrasonic assembly: A method of assembling two or more thermoplastic parts by welding the mating surfaces through application of vibratory mechanical pressure at ultrasonic frequencies at the interfaces that melt and fuse together.

Ultrasonic insertion: A method of inserting metal parts into thermoplastic moldings by applying vibratory mechanical pressure at ultrasonic frequencies to the metal parts, and thereby to the thermoplastic. The thermoplastic melts at the interface between it and the insert, and remolds itself around the part.

1.3 Other Terms

Engineering plastics: Plastics materials normally characterized by a high order of performance characteristics that are well defined and predictable. They have also been described as having long-term heat stability, high heat distortion temperature, high tensile and flexural characteristics, good impact strength, low coefficient of friction, excellent abrasion resistance, and high tolerance to hydrocarbon solvents.

Plastic (A): Pliable and capable of being shaped by pressure.

Plastics (N): A generic term for the industry and its products that include polymeric substances, natural or synthetic, excluding rubbers.

References

1. Extracted from AGMA 141.01 as provided for informational purposes only and is not to be construed to be a part of American Gear Manufacturers Association Information Sheet 141.01, *Plastics Gearing — Molded and Machined, and Other Methods, a Report on the State of the Art*.

2. M. Grayson, Ed., *Encyclopedia of Composite Materials and Components*, John Wiley and Sons, New York (1983).

3. M. and I. Ash, Eds., *Encyclopedia of Plastics, Polymers and Resins*, 3 volumes, Chemical Publishing Co., New York (1982).

4. L. R. Whittington, *Whittington's Dictionary of Plastics*, Technomic Publishing Co., Lancaster, PA (1978).

5. E. Miller, Ed., *Plastics Products Design Handbook, Part A: Materials and Components*, Marcel Dekker, New York (1981).

6. E. Miller, Ed., *Plastics Products Design Handbook, Part B: Processes and Design for Processes*, Marcel Dekker, New York (1983).

Appendix D

Equations for Tolerances

1. Introduction

The tolerances in Tables 7.9 and 7.10 can be obtained from the follow-ing equations. They are provided for those who wish to utilize stored program computers for gear calculations.

2. Equations

Symbols used in these equations are as follows:

Pd = diametral pitch
D = pitch diameter
Qn = quality number
F = face width

All values are in ten-thousandths of an inch.

1. Runout tolerance = $58(D)^{.238}(P_d)^{-.484}(1.4)^{8-Q_n}$

2. Pitch tolerance = $10.5(D)^{.177}(P_d)^{-.244}(1.42)^{8-Q_n}$

3. Profile tolerance = $21.5(D)^{.154}(P_d)^{-.435}(1.4)^{8-Q_n}$

    a. For face width 1 inch or less:

    Lead tolerance = $(-.00244(Q_n)^3 + .13638(Q_n)^2 - 2.69177(Q_n)$
    $+ 18.955)$

379

b. For face width greater than 1 inch:

Lead tolerance = $(-.00244(Q_n)^3 + .13638(Q_n)^2 - 2.69177(Q_n) + 18.955)$ $(F)^{.72}$

4. Tooth-to-tooth composite tolerance

a. $P_d \times D$ equal or less than 20

Tolerance = $54.7(P_d)^{-.48}(D)^{-.24}(1.4)^{8-Q_n}$

b. $P_d \times D$ greater than 20 but equal or less than 32

Tolerance = $38.2(P_d)^{-.36}(D)^{-.13}(1.4)^{8-Q_n}$

c. $P_d \times D$ greater than 32

Tolerance = $25(P_d)^{-.24}(1.4)^{8-Q_n}$

5. Total composite tolerance

a. For $P_d \times D$ equal or less than 20.2

$$\text{Tolerance} = 15\left((20.2/P_d)^{(.24)}(P_d)^{-.15}\right)$$
$$\left(1.16^{(10-Y)}(1.4)^{(8-Q_n)}-(.075)\right)$$
$$(20-P_d)\left((20/P_D)-D\right)$$

b. For $P_d \times D$ greater than 20.2

$$\text{Tolerance} = 14.5\left((D)^{.24}(P_d)^{-.15}\right)(1.16)^{10-Y}(1.4)^{8-Q_n}$$

where

$$Y = (5.0337 \log_{10}(P_d))-.5153$$

Source: Gear Handbook, Vol. 1: Gear Classification, Materials, and Measuring Methods of Unassembled Gears (AGMA 390.03) American Gear Manufacturers Association, Alexandria, VA (1973).

Index